AMERICAN MATHEMATI
COLLOQUIUM PUBLICATION

THE PRINCETON COLLOQUIUM
1909

PART I

FUNDAMENTAL EXISTENCE
THEOREMS

BY

GILBERT AMES BLISS

NEW YORK

PUBLISHED BY THE

AMERICAN MATHEMATICAL SOCIETY

501 WEST 116TH STREET

NEW YORK CITY

1913

REPRINTED 1934

<u>Printing Statement:</u>

Due to the very old age and scarcity of this book,
many of the pages may be hard to read due to the
blurring of the original text, possible missing pages,
missing text, dark backgrounds and other issues
beyond our control.

Because this is such an important and rare work, we
believe it is best to reproduce this book regardless of
its original condition.

Thank you for your understanding.

CONTENTS

i

CHAPTER III

EXISTENCE THEOREMS FOR DIFFERENTIAL EQUATIONS

FUNDAMENTAL EXISTENCE THEOREMS

GILBERT AMES BLISS

INTRODUCTION

The existence theorems to which these lectures are devoted have been the subject of a long sequence of investigations extending from the time of Cauchy to the present day, and have found application at the basis of a variety of mathematical theories including, as perhaps of especial importance, the theory of algebraic functions and the calculus of variations. If a single solution $(a; b) = (a_1, a_2, \cdots, a_m; b_1, b_2, \cdots, b_n)$ of a set of equations

$$f_a(x_1, x_2, \cdots, x_m; y_1, y_2, \cdots, y_n) = 0 \quad (\alpha = 1, 2, \cdots, n)$$

is known, then in a neighborhood of $(a ; b)$ there is one and only one other solution corresponding to each set of values x in a properly chosen neighborhood of the values a, and in the totality of solutions $(x ; y)$ so defined the variables y are single-valued and continuous functions of the x's. If a set of initial constants $(\xi, \eta_1, \eta_2, \cdots, \eta_n)$ is given, then in a neighborhood of these values there is one and but one continuous arc

$$y_a = y_a(x) \qquad (\alpha = 1, 2, \cdots, n)$$

satisfying the differential equations

$$\frac{dy_a}{dx} = g_a(x, y_1, y_2, \cdots, y_n) \quad (\alpha = 1, 2, \cdots, n)$$

and passing through the initial values η when $x = \xi$.

2   1

The formulation and first satisfactory proofs of these theorems, at least for the case where only two variables x, y are involved, seem to be ascribed with unanimity to Cauchy. For the implicit functions his proof rested upon the assumption that the function f should be expressible by means of a power series, and the solution he sought was also so expressible, a restriction which was later removed with remarkable insight by Dini. For a differential equation, on the other hand, Cauchy assumed only the continuity of the function g and its first derivative for y, and his method of proof, with the well-known alteration due to Lipschitz, retains to-day recognized advantages over those of later writers.

In the following pages (§§ 1, 16) the two theorems stated above are proved with such alterations in the usual methods as seemed desirable or advantageous in the present connection. The proof given for the fundamental theorem of implicit functions is applicable when the independent variables x are replaced by a variable p which has a range of much more general type than a set of points in an m-dimensional x-space.* It is not necessary always to know an initial solution in order that others may be found. In the treatment of Kepler's equation, for example, which defines the eccentric anomaly of a planet moving in an elliptical orbit in terms of the observed mean anomaly, one starts with an approximate solution only and determines an exact solution by means of a convergent succession of approximations. This procedure is closely allied to a method of approximation due to Goursat (§ 3), suggested apparently by Picard's treatment of the existence theorem for differential equations.

One of the principal purposes of the paragraphs which follow, however, is to free the existence theorems as far as possible from

---

* The notion of a general range has been elucidated by Moore, The New Haven Mathematical Colloquium, page 4, the special cases which he particularly considers being enumerated on page 13. An application of the method of § 1 of these lectures when the range of p is a set of continuous curves, has been made by Fischer, "A generalization of Volterra's derivative of a function of a line," Dissertation, Chicago (1912).

the often inconvenient restriction which is implied by the words " in a neighborhood of," or which is so aptly expressed in German by the phrase "im Kleinen." It is evident from very simple examples that the totality of solutions $(x; y)$ associated continuously with a given initial solution of a system of equations $f = 0$ of the form described above, can not in general have the property that the variables y are everywhere single-valued functions of the variables x, and the result of attempting, perhaps unconsciously, to preserve the single-valued character of the solutions has been the restriction of the region to which the existence theorems apply. In order to avoid this difficulty and to characterize to some extent the totality of solutions associated continuously with a given initial one in a region specified in advance, the writer has introduced (§ 5) the notion of a particular kind of point set called a sheet of points. In a suitably chosen neighborhood of a point $(a; b)$ of the sheet there corresponds to every set of values x sufficiently near to the values a exactly one point $(x; y)$ of the sheet, and the single-valued functions y so determined are continuous and have continuous first derivatives. This condition does not at all imply that there are no other points of the sheet outside the specified neighborhood of the point $(a; b)$ and having a projection x near to a. With the help of the notion of a sheet of points it can be concluded that with any initial solution $(a; b)$ of the equations $f = 0$ there is associated a unique sheet S of solutions whose only boundary points are so-called exceptional points where the functions f either actually fail, or else are not assumed, to have the continuity and other properties which are demanded in the proof of the well-known theorem for the existence of solutions in a neighborhood of an initial one. It is important oftentimes to know whether or not a sheet of solutions is actually single-valued throughout its entire extent, and a criterion sufficient to ensure this property has also been derived (§ 7).

On the basis of these results some important theorems concerning the transformation of plane regions into regions of another plane by means of equations of the form

$$x_1 = \psi_1(y_1, y_2), \qquad x_2 = \psi_2(y_1, y_2),$$

as in the theory of conformal transformation, have been deduced
(§ 8). If the functions ψ have suitable continuity properties
and a non-vanishing functional determinant in the interior of a
simply closed regular curve B in the y-plane, and if B is transformed into a simply closed regular curve A of the x-plane, then
the equations define a one-to-one correspondence between the
interiors of A and B, and the inverse functions so defined have
continuity properties similar to those of ψ_1 and ψ_2. This is but
a sample of the theorems which may be stated. Others are also
given (§ 8) which apply to the transformation of regions not
necessarily finite, and to systems containing more than two
equations.

The theory of the singularities of implicit functions is of con-
siderable difficulty and has been but incompletely developed.
For a transformation of the form above in which the functions
ψ_1, ψ_2 are analytic, the singular point to be studied, at which the
functional determinant $D = \partial(\psi_1, \psi_2)/\partial(y_1, y_2)$ vanishes, as
well as its image in the x-plane, may both without loss of generality be supposed at the origin. The most general case under
these circumstances is that for which the determinant D does
not vanish identically and the equations $\psi_1 = 0$, $\psi_2 = 0$ have no
real solutions in common near the origin except the values
$y_1 = y_2 = 0$ themselves. It is found that the branches of the
curve $D = 0$ bound off with a suitably chosen circle about the
origin a number of triangular regions. Each of these regions is
transformed in a one-to-one way into a sort of Riemann surface
on the x-plane which winds about the origin and is bounded by
the image of the boundary of the triangular region (see § 11,
Fig. 6). If the signs of D in two adjacent triangular regions
are opposite, then their images overlap along the common
boundary; otherwise they adjoin without overlapping. At any
point of one of the Riemann surfaces the inverse functions defined

by the transformation are continuous and in the interior of the surface they have everywhere continuous derivatives. These results are obtained by means of applications of the theorem described above for the transformation of the interior of a simply closed curve B; and the same method of procedure would undoubtedly be of service when the curves $\psi_1 = 0$, $\psi_2 = 0$ have real branches through the origin in common, which must occur whenever they have common points in every neighborhood of the values $y_1 = y_2 = 0$. The case where the determinant D vanishes identically is also considered (§ 12).

For the singularities of implicit functions defined by a system of equations $f = 0$ there is a generalization of the preparation theorem of Weierstrass (§ 9) suggested to the writer by some remarks in the introduction of Poincaré's Thesis, and by a study of the elimination theory of Kronecker for algebraic equations. The theorem is presented here (§ 13) for two equations and two variables y_1, y_2 in the form originally given at the time of the Princeton Colloquium, but the method of proof is similar to that of a later paper* and applies with suitable modifications to a system containing more equations and independent variables. These results can not by any means be said to afford a complete characterization of the singularities of implicit functions, but it is hoped that they may be useful in paving the way for researches of a more comprehensive character.

The writer published some years ago a paper† concerning the extensibility of the solutions of a system of differential equations, of the form specified above, from boundary to boundary of a finite closed region R in which the functions g_a are supposed to have suitable continuity properties. In the last chapter of these lectures the character of the region has been generalized so that no restrictions as to its finiteness or closure are made, and it is shown that the approximations of Cauchy converge to a solution over an interval

---

* See the footnote to page 73.

† " The solutions of differential equations of the first order as functions of their initial values," *Annals of Mathematics*, 2d series, vol. 6 (1904), page 49.

in the interior of which the limiting curve is continuous and
interior to R, while at the ends of the interval the only limit
points of the curve are at infinity or else are on the boundary of the
region.   The solutions so defined are continuous and differentiable with respect to their initial values, a property which once
proved is of great service in many of the applications of the
existence theorems.   One situation in which these results have
an important bearing is related to a partial differential equation
of the first order

$$F(x, y, z, \partial z/\partial x, \partial z/\partial y) = 0.$$

When this equation is analytic, any analytic curve C, which is
not a so-called integral curve, defines uniquely an analytic surface
containing the curve and satisfying the differential equation.   The
uniqueness in this case is a consequence, in the first place, of
the fact that an analytic surface is completely determined when
an initial series defining its values in a limited region is given,
and, in the second place, of the theorem that at a given point
and normal of the initial curve C satisfying the differential equation there is but one series defining an integral surface including
the points of C and having the given initial normal.   It is not
self evident in what sense a solution of a non-analytic equation
is uniquely determined by an initial curve, as may be seen by very
simple examples.   An initial curve which is not an integral curve
will in general have associated with it, however, a strip of normals which satisfy the partial differential equation, and whose
elements as initial values determine a one-parameter family of
characteristic strips simply covering a region R_{xy} of the xy-plane
about the projection of the initial curve C.   There is one and but
one integral surface of the differential equation with a continuously turning tangent plane and continuous curvature, which is
defined at every point of the region R_{xy} and contains the initial
curve C and its strip of normals (§ 19).

CHAPTER I

ORDINARY POINTS OF IMPLICIT FUNCTIONS

§1. The Fundamental Theorem

The fundamental theorem of the implicit function theory states the existence of a set of functions

$$y_\alpha = y_\alpha(x_1, x_2, \cdots, x_m) \cdot (\alpha = 1, 2, \cdots, n)$$

which satisfy a system of equations of the form

$$(1) \qquad f_\alpha(x_1, x_2, \cdots, x_m; y_1, y_2, \cdots, y_n) = 0 \quad (\alpha = 1, 2, \cdots, n)$$

in a neighborhood of a given initial solution $(a; b)$. Dini's method,[*] for the case in which the functions f are only assumed to be continuous and to have continuous first derivatives, is to show the existence of a solution of a single equation, and then to extend his result by mathematical induction to a system of the form given above, a plan which has been followed, with only slight alterations and improvements in form, by most writers on the theory of functions of a real variable. In a more recent paper[†] Goursat has applied a method of successive approximations which enabled him to do away with the assumption of the existence of the derivatives of the functions f with respect to the independent variables x.

One can hardly be dissatisfied with either of these methods of attack. It is true that when the theorem is stated as precisely as in the following paragraphs, the determination of the neighborhoods at the stage when the induction must be made is rather inelegant, but the difficulties encountered are not serious. The introduction of successive approximations is an interesting step,

---

[*] Lezioni di Analisi infinitesimale, vol. 1, chap. 13. For historical remarks, see Osgood, Encyclopädie der mathematischen Wissenschaften, II, B 1, § 44 and footnote 30.

[†] Bulletin de la Société mathématique de France, vol. 31 (1903), page 185.

though it does not simplify the situation and indeed does not add generality with regard to the assumptions on the functions f. The method of Dini can in fact, by only a slight modification, be made to apply to cases where the functions do not have derivatives with respect to the variables x. The proof which is given in the following paragraphs seems to have advantages in the matter of simplicity over either of the others. It applies equally well, without induction, to one or a system of equations, and requires only the initial assumptions which Goursat mentions in his paper.

Where it is possible without sacrificing clearness, the row letters f, x, y, p, a, b will be used to denote the systems

$$
\begin{aligned}
&f = (f_1, f_2, \cdots, f_n), \qquad x = (x_1, x_2, \cdots, x_m), \\
&y = (y_1, y_2, \cdots, y_n), \qquad a = (a_1, a_2, \cdots, a_m), \\
&b = (b_1, b_2, \cdots, b_n), \qquad p = (a_1, a_2, \cdots, a_m; b_1, b_2, \cdots, b_n).
\end{aligned}
$$

In this notation the equations (1) have the form

$$f(x; y) = 0,$$

the interpretation being that every element of f is a function of $x_1, x_2, \cdots, x_m; y_1, y_2, \cdots, y_n$, and every f_i is to be set equal to zero. The notations p_ϵ, a_ϵ, b_ϵ represent respectively the neighborhoods

$$|x - a| < \epsilon, \quad |y - b| < \epsilon; \quad |x - a| < \epsilon; \quad |y - b| < \epsilon$$

of the points p, a, b.

With these notations in mind the fundamental theorem which is to be proved may be stated as follows:

Hypotheses:

1) *the functions $f(x; y)$ are continuous, and have first partial derivatives with respect to the variables y which are also continuous, in a neighborhood of the point $(a; b)$ which will be denoted by p;*

2) $f(a; b) = 0$;

3) *the functional determinant $D = \partial(f_1, f_2, \cdots, f_n)/\partial(y_1, y_2, \cdots, y_n)$ is different from zero at p.*

Conclusions.

1) *a neighborhood p_ϵ can be found in which there corresponds to a given value x at most one solution $(x; y)$ of the equations $f(x; y) = 0$;*

2) *for any neighborhood p_ϵ with the property just described a constant $\delta \leqq \epsilon$ can be found such that every x in a_δ has associated with it a point $(x; y)$ which satisfies the equations $f(x; y) = 0$;*

3) *the functions $y(x_1, x_2, \cdots, x_m)$ so found are continuous in the region a_δ.*

For the neighborhood p_ϵ let one be chosen in which the continuity properties of the functions f are preserved. If $(x; y)$ and $(x; y')$ are two points in p_ϵ, it follows, by applying Taylor's formula to the differences $f(x; y') - f(x; y)$, that

$$f_1(x; y') - f_1(x; y) = \frac{\partial f_1}{\partial y_1}(y_1' - y_1) + \cdots + \frac{\partial f_1}{\partial y_n}(y_n' - y_n),$$

$$\cdot \quad \cdot \quad \cdot \quad \cdot \quad \cdot \quad \cdot \quad \cdot \quad \cdot \quad \cdot \quad \cdot$$

$$f_n(x; y') - f_n(x; y) = \frac{\partial f_n}{\partial y_1}(y_1' - y_1) + \cdots + \frac{\partial f_n}{\partial y_n}(y_n' - y_n),$$

where the arguments of the derivatives $\partial f_\alpha / \partial y_\beta$ have the form $x; y + \theta_\alpha(y' - y)$, and $0 < \theta_\alpha < 1$. The determinant of these derivatives is different from zero when $(x; y') = (x; y) = (a; b)$, and hence must remain different from zero if p_ϵ is restricted so that in it the functional determinant D remains different from zero. It is then impossible that $(x; y)$ and $(x; y')$ should both be solutions of the equations $f(x; y) = 0$, if y is distinct from y'.

In the corresponding region b_ϵ the function

$$\varphi(a; y) = f_1^2(a; y) + f_2^2(a; y) + \cdots + f_n^2(a; y)$$

has a minimum for $y = b$, since for that value it vanishes and for every other it is positive. In particular

$$\varphi(a; \eta) - \varphi(a; b) > m > 0$$

when η ranges over the *closed* set of points η forming the boundary of b_ϵ, on account of the continuity of φ, and the inequality

$$\varphi(x; \eta) - \varphi(x; b) > m$$

remains true for all values x in a suitably chosen domain a_δ. Hence for a fixed x in a_δ the minimum of $\varphi(x; y)$ is attained at a point y *interior* to b_ϵ. At such a point, however,

$$\frac{1}{2}\frac{\partial \varphi}{\partial y_1} = f_1 \frac{\partial f_1}{\partial y_1} + f_2 \frac{\partial f_2}{\partial y_1} + \cdots + f_n \frac{\partial f_n}{\partial y_1} = 0,$$

$$\cdot \qquad \cdot \qquad \cdot$$

$$\frac{1}{2}\frac{\partial \varphi}{\partial y_n} = f_1 \frac{\partial f_1}{\partial y_n} + f_2 \frac{\partial f_2}{\partial y_n} + \cdots + f_n \frac{\partial f_n}{\partial y_n} = 0,$$

and this can happen only when all the elements of f are zero, since the functional determinant D is different from zero in p_ϵ. It follows that to every point x in a_δ there corresponds in p_ϵ a solution $(x; y)$ of the equations $f(x; y) = 0$.

The functions $y(x_1, x_2, \cdots, x_m)$ defined in this way over the region a_δ are all continuous. For consider the values y and $y + \Delta y$ corresponding to two points x and $x + \Delta x$. By applyng Taylor's formula it follows from the relations

$$f(x; y + \Delta y) - f(x; y) = f(x; y + \Delta y) - f(x + \Delta x; y + \Delta y),$$

which are true because $(x; y)$ and $(x + \Delta x; y + \Delta y)$ both make $f = 0$, that

$$\frac{\partial f_1}{\partial y_1}\Delta y_1 + \frac{\partial f_1}{\partial y_2}\Delta y_2 + \cdots + \frac{\partial f_1}{\partial y_n}\Delta y_n$$
$$= f_1(x; y + \Delta y) - f_1(x + \Delta x; y + \Delta y),$$

(2) $\qquad \cdot \qquad \cdot \qquad \cdot \qquad \cdot \qquad \cdot \qquad \cdot \qquad \cdot$

$$\frac{\partial f_n}{\partial y_1}\Delta y_1 + \frac{\partial f_n}{\partial y_2}\Delta y_2 + \cdots + \frac{\partial f_n}{\partial y_n}\Delta y_n$$
$$= f_n(x; y + \Delta y) - f_n(x + \Delta x; y + \Delta y),$$

where the arguments of the derivatives $\partial f_\alpha / \partial y_\beta$ have the form $x; y + \theta_\alpha \Delta y \ (0 < \theta_\alpha < 1)$. The determinant of these derivatives is different from zero on account of the way in which p_ϵ was chosen, and the second members of the equations approach zero with Δx. Hence the same must be true of the quantities

Δy, and thus the functions $y(x_1, x_2, \cdots, x_m)$ are seen to be continuous.

A similar application of Taylor's formula leads to the conclusion:

If the functions f have derivatives of the first order with respect to x_k which are continuous in the neighborhood of p, so have also the functions $y(x_1, x_2, \cdots, x_m)$ in the region a_δ; and if the f's have all derivatives of the nth order continuous, so have the functions $y(x_1, x_2, \cdots, x_m)$.

For suppose

$$\Delta x_1 \neq 0, \quad \Delta x_2 = \Delta x_3 = \cdots = \Delta x_m = 0.$$

Then by applying Taylor's formula to the second members of equations (2) it follows that

$$\frac{\partial f_1}{\partial y_1}\frac{\Delta y_1}{\Delta x_1} + \frac{\partial f_1}{\partial y_2}\frac{\Delta y_2}{\Delta x_1} + \cdots + \frac{\partial f_1}{\partial y_n}\frac{\Delta y_n}{\Delta x_1} + \frac{\partial f_1}{\partial x_1} = 0,$$

$$\cdots \cdots \cdots \cdots \cdots \cdots$$

$$\frac{\partial f_n}{\partial y_1}\frac{\Delta y_1}{\Delta x_1} + \frac{\partial f_n}{\partial y_2}\frac{\Delta y_2}{\Delta x_1} + \cdots + \frac{\partial f_n}{\partial y_n}\frac{\Delta y_n}{\Delta x_1} + \frac{\partial f_n}{\partial x_1} = 0,$$

where the arguments of the derivatives $\partial f_\alpha / \partial x_1$ have the form $x + \theta_\alpha'\Delta x; y + \Delta y$. Hence as Δx_1 approaches zero the quotients $\Delta y_\alpha / \Delta x_1$ approach limits $\partial y_\alpha / \partial x_1$ which satisfy the equations

$$\frac{\partial f_1}{\partial y_1}\frac{\partial y_1}{\partial x_1} + \frac{\partial f_1}{\partial y_2}\frac{\partial y_2}{\partial x_1} + \cdots + \frac{\partial f_1}{\partial y_n}\frac{\partial y_n}{\partial x_1} + \frac{\partial f_1}{\partial x_1} = 0,$$

(3) $$\cdots \cdots \cdots \cdots \cdots \cdots$$

$$\frac{\partial f_n}{\partial y_1}\frac{\partial y_1}{\partial x_1} + \frac{\partial f_n}{\partial y_2}\frac{\partial y_2}{\partial x_1} + \cdots + \frac{\partial f_n}{\partial y_n}\frac{\partial y_n}{\partial x_1} + \frac{\partial f_n}{\partial x_1} = 0,$$

where the arguments of the derivatives of f are now $(x; y)$. A similar consideration shows the existence of the first derivatives with respect to the variables $x_2, x_3, \cdots, x_m$. The existence of the higher derivatives follows from the observation that the solutions of equations (3) for the quotients $\partial f_\alpha / \partial y_\beta$ are differentiable $n - 1$ times with respect to the variables x, on account of the assumption that the functions f are differentiable n times.

§ 2. EQUATIONS IN WHICH THE FUNCTIONS ARE ANALYTIC

It seems necessary to proceed differently in order to prove that when the functions f in equations (1) are analytic with coefficients and variables permitted to assume imaginary values, the solutions $y = y(x_1, x_2, \cdots, x_m)$ are also analytic functions of the variables x. The following theorem can first be proved:

When the functions f are formal series in the variables x; y with literal coefficients and having no constant terms, then there exists one and but one set of series

(4) $$y_a = y_a(x_1, x_2, \cdots, x_m)$$

for the variables y, which vanish with the x s and satisfy identically the equations $f(x; y) = 0$. Each coefficient in the series y is rational in a finite number of those of the functions f, the only denominators occurring being powers of the determinant R of the coefficients of the linear terms in y.

To prove this let the equations $f = 0$ be written in the form

$$a_{11}y_1 + a_{12}y_2 + \cdots + a_{1n}y_n = g_1(x; y),$$
$$\cdot \qquad \cdot \qquad \cdot \qquad \cdot \qquad \cdot \qquad \cdot \qquad \cdot$$
$$a_{n1}y_1 + a_{n2}y_2 + \cdots + a_{nn}y_n = g_n(x; y),$$

where the functions g have no linear terms in y. By multiplying these equations by proper factors and adding, they may be made to take the form

(5) $$y_a = h_a(x; y) \qquad (\alpha = 1, 2, \cdots, n),$$

where the series h have still no linear terms in y and have coefficients which are rational in those of the functions f, the only denominators occurring being the determinant R. Any series for y which satisfy formally the original equations must satisfy the last equations, and vice versa.

Consider now a set of series (4) in which the coefficients are indeterminates c. If they satisfy the equations (5) identically, then by comparison of coefficients on the two sides it is seen that any coefficient c_ν of a term of degree ν must be equal to a polynomial, with positive integral coefficients, in a finite number of the coefficients of the functions h and in the coefficients $c_{\nu-k}$ of terms in the functions y of lower degree than ν. For there are at most a finite number of terms on the right of any given degree ν, and since the functions h have no linear terms in the variables y it follows that wherever the term containing c_ν occurs it is always multiplied by a y or by a power of some of the variables x, and hence c_ν can only appear in terms of degree greater than ν. Since the coefficients of the linear terms in the functions y are equal respectively to corresponding coefficients in the functions h, it follows by an easy induction that every coefficient in the functions y must be a polynomial with positive integral coefficients in a finite number of the coefficients of the functions h. There is evidently but one set of series (4) of the kind described satisfying formally the equations (5), or what is the same thing, the equations $f = 0$.

For any numerical choice of the coefficients of the functions f in the domain of real or imaginary numbers for which the series f converge and the determinant $R = |\,a_{\alpha\beta}\,|$ is different from zero, the series (4) for y will also be well-determined and convergent.

For, a set of equations

(6) $$y_\alpha = H_\alpha(x; y) \qquad (\alpha = 1, 2, \cdots, n)$$

can be constructed whose coefficients are all positive and greater numerically than the corresponding coefficients in the functions h, and for which the corresponding series $y = Y(x_1, x_2, \cdots, x_m)$ converge. The coefficients in the functions Y will be greater numerically than the corresponding coefficients of the series $y(x_1, x_2, \cdots, x_m)$, and hence the series y will also converge.

To show this suppose that ρ is a positive constant smaller than the radii of convergence of the functions $h(x; y)$. Then the series $h(\rho; \rho)$ are convergent, and each term is numerically smaller than a constant M chosen greater than the sum of the absolute values of the terms in any one of the series $h(\rho; \rho)$. The coefficient of any term in $h(x; y)$ is less than M/ρ^ν where ν is the degree of the term. The series

$$H_a(x; y) = \cfrac{M}{\left(1 - \cfrac{x_1 + x_2 + \cdots + x_m}{\rho}\right)\left(1 - \cfrac{y_1 + y_2 + \cdots + y_n}{\rho}\right)} \\ - M - M\frac{y_1 + y_2 + \cdots + y_n}{\rho}$$

are similar to the series $h(x; y)$ in the matter of missing terms, and dominate them in the manner described above, since the coefficient of any term of degree ν is M/ρ^ν or greater.

The unique series satisfying equations (6) will evidently be convergent if a convergent series u in x can be determined satisfying

$$u = \cfrac{M}{\left(1 - \cfrac{x_1 + x_2 + \cdots + x_m}{\rho}\right)\left(1 - \cfrac{nu}{\rho}\right)} - M - M\frac{nu}{\rho},$$

for then every series y can be put equal to that series u. The latter equation is however a quadratic in u and has the solution

$$u = \frac{\rho^2}{2n(\rho + Mn)}\left\{1 - \sqrt{1 - \frac{4Mn(\rho + Mn)}{\rho^2}\cdot\cfrac{\dfrac{x_1 + x_2 + \cdots + x_m}{\rho}}{1 - \dfrac{x_1 + x_2 + \cdots + x_m}{\rho}}}\right\}$$

vanishing with x. This will certainly be representable by a convergent series in x provided that

$$|x_i| < \frac{\rho^3}{m(\rho + 2Mn)^2} \qquad (i = 1, 2, \cdots, m),$$

since then the second term under the radical is numerically less than unity.

The two theorems which have just been proved enable one to make the following statement concerning the solutions whose existence was proved in § 1:

If the functions $f(x; y)$ are analytic in the region p_ϵ, then the solutions (4) of the equations $f(x; y) = 0$ are analytic at every point of the region a_δ.

It is only necessary to transform the origin of coordinates to the particular point $(x; y)$ of the solution which it is desired to investigate.

Furthermore when the domain in which the equations $f = 0$ are to be studied is the domain of complex numbers, a theorem analogous to that of § 1 may be stated.

If in the domain of complex numbers the functions $f(x; y)$ are analytic at a point $p(a; b)$ at which

$$f(a; b) = 0, \quad D(a; b) = \left[\frac{\partial(f_1, f_2, \cdots, f_n)}{\partial(y_1, y_2, \cdots, y_n)} \right]_{\substack{x=a \\ y=b}} \neq 0,$$

then there exists a neighborhood p_ϵ in which any x corresponds to at most one solution $(x; y)$, either real or complex, of the equations $f(x; y) = 0$. For any such choice of p_ϵ a neighborhood a_δ ($\delta \leqq \epsilon$) can be found such that every point x in a_δ has associated with it a solution $(x; y)$ of the equations $f = 0$ in p_ϵ, and the values y for these solutions are defined by a set of functions

$$(7) \qquad\qquad y_\alpha = y_\alpha(x_1, x_2, \cdots, x_m) \qquad (\alpha = 1, 2, \cdots, n)$$

which are expressible as series in the differences $x - a$ convergent in the region a_δ.

The existence of the neighborhood p_ϵ is provable by the argument used in § 1, since for any two points $(x; y)$ and $(x; y')$ in the common domain of convergence of the functions f, equations of the form

$$f_\alpha(x; y') - f_\alpha(x; y) = A_{\alpha 1}(y_1' - y_1) + \cdots + A_{\alpha n}(y_n' - y_n),$$
$$(\alpha = 1, 2, \cdots, n)$$

hold, where the coefficient $A_{\alpha\beta}$ is a convergent series in the differences $x - a$, $y - b$, $y' - b$ with constant term equal to $a_{a\beta}$. The existence of the coefficients A can be established by considering two analogous terms in $f(x; y)$ and $f(x; y')$. The difference of such a pair of terms will always be linearly expressible in terms of the differences

$$(y_a' - b_a) - (y_a - b_a) = y_a' - y_a \quad (\alpha = 1, 2, \cdots, n).$$

Furthermore for $(x, y, y') = (a, b, b)$ the derivative of the first member with respect to y_β' reduces to $a_{a\beta}$, while that of the second is the constant term in $A_{a\beta}$. Hence for these values of the variables the determinant $| A_{a\beta} |$ reduces to $D(a, b) \neq 0$.

By transforming the origin of coordinates to the point (a, b) and applying the first two theorems of this section, it follows that there exists a set of convergent series (7) satisfying the equations $f = 0$ identically; and for a sufficiently small region a_δ the points $(x; y)$ which they define will all lie in the neighborhood p_ϵ.

§ 3. Goursat's Method of Approximation

The method of approximation which is to be presented in the following paragraphs is of interest primarily because it affords a direct method of finding the values of implicit functions, and justifies computations sometimes used in the applications of the theory. In order to exhibit this method suppose again that the functions f have the properties described in the principal theorem of § 1, and consider the following set of equations suggested by Taylor's formula·

$$f_1(x; y) + a_{11}(y_1' - y_1) + a_{12}(y_2' - y_2) + \cdots \\ + a_{1n}(y_n' - y_n) = 0,$$

(8)

$$f_n(x; y) + a_{n1}(y_1' - y_1) + a_{n2}(y_2' - y_2) + \cdots \\ + a_{nn}(y_n' - y_n) = 0,$$

in which the coefficient $a_{a\beta}$ is the value of $\partial f_a/\partial y_\beta$ at the point p. When solved for the variables y', these equations take the form

(9) $$y_a' = \varphi_a'(x; y) \quad (\alpha = 1, 2, \cdots, n),$$

and one verifies readily by substitution of these expressions in equations (8) that the functions φ and all of their first derivatives with respect to the elements of y are continuous near p; and at the point p itself φ_a has the value b_a, while all of its derivatives with respect to the y's vanish.

A sequence of systems $y^{(k)} = (y_1^{(k)}, y_2^{(k)}, \cdots, y_n^{(k)})$ beginning with the set

$$y' = [\varphi_1(x; b), \varphi_2(x; b), \cdots, \varphi_n(x; b)]$$

can now be defined by means of the recursion formulas (9), which are equivalent to

$$y_a^{(k)} = \varphi_a(x; y^{(k-1)}) \qquad (\alpha = 1, 2, \cdots, n).$$

Let p_ϵ be any neighborhood of p in which the continuity properties of f are retained, and in which the derivatives of φ remain numerically less than θ/n where $0 < \theta < 1$. If the values of x are restricted to a region $a_\delta (\delta \leqq \epsilon)$ so small that every element of the set y' satisfies the inequality

$$(10) \qquad | y_a' - b_a | < \epsilon(1 - \theta),$$

then the points $(x; y^{(k)})$ will all lie in the neighborhhood p_ϵ and will approach uniformly a limiting point $(x; y)$ which is a solution of the equations (1).

To prove these statements one needs only to apply successively the inequality

$$| y_a^{(k)} - y_a^{(k-1)} | = | \varphi_a(x; y^{(k-1)}) - \varphi_a(x; y^{(k-2)}) |$$
$$\leqq \frac{\theta}{n}\{ | y_1^{(k-1)} - y_1^{(k-2)} | + | y_2^{(k-1)} - y_2^{(k-2)} | $$
$$+ \cdots + | y_n^{(k-1)} - y_n^{(k-2)} |\},$$

which follows readily by an application of Taylor's formula. Since the inequalities (10) hold, the last formula successively applied shows that

$$| y_a^{(k)} - y_a^{(k-1)} | \leqq \theta^{k-1}\epsilon(1 - \theta).$$

Consequently the sum $y_a^{(k)}$ of the first $k + 1$ terms of the series

$$(11) \quad b_a + (y_a' - b_a) + (y_a'' - y_a') + \cdots + (y_a^{(k)} - y_a^{(k-1)}) + \cdots$$

3

differs in absolute value from b_a by a quantity which is less than

$$\epsilon(1 - \theta)(1 + \theta + \theta^2 + \cdots + \theta^{k-1}) = \epsilon(1 - \theta^k) < \epsilon.$$

Hence the points $(x; y)$ all lie in the neighborhood p_ϵ, and the series (11) is uniformly convergent in the neighborhood a_δ.

The limiting point $(x; y)$ evidently satisfies the equations $f = 0$. For at every stage the values $(x, y, y') = (x, y^{(k-1)}, y^{(k)})$ satisfy the equations (8), and the first members of these equations approach uniformly the values $f(x; y)$.

The process of determining the solutions described above is evidently one of trial and error. The values $y = b$ being first substituted, the equations (9) determine approximately the correction $y' - b$ which must be added to b in order to obtain a solution for any value of x near to a. For the values so corrected the equations (9) give again a new correction $y'' - y'$, and so on.

It is ordinarily presupposed that an initial solution $(a; b)$ is given, *but the process may also lead to the discovery of a solution in case only an initial point which approximately satisfies the equation is known.* To show this suppose that the functions f are continuous and have continuous first partial derivatives with respect to the variables y in a closed region R of points $(x; y)$ in which the functional determinant $D(x; y)$ is different from zero. The functions φ in equations (9) are to be thought of as depending upon $(x; y)$, and also upon the variables $(a; b)$ which enter in the derivatives $a_{\alpha\beta}$. Then the expressions $\varphi(x, y, a, b)$, $\varphi_y(x, y, a, b)$ are continuous when $(x; y)$, $(a; b)$ lie in R, and all of the derivatives φ_y vanish identically when $(x; y) = (a; b)$. The value of $\varphi(a, b, a, b)$ is not necessarily b, however, when $(a; b)$ is not a solution. Two positive constants, $\theta < 1$ and ϵ, can be determined so that

$$|\,\varphi_y(x, y, a, b)\,| < \theta/n$$

whenever $(a; b)$ and $(x; y)$ satisfy the inequalities

$$|\,x - a\,| < \epsilon, \qquad |\,y - b\,| < \epsilon.$$

If now there exists a point $p(a; b)$ for which the neighborhood p_ϵ

is entirely within R, and such that

$$| \varphi(a, b, a, b) - b | < \epsilon(1 - \theta),$$

then the sequence $y^{(k)}$ defined converges uniformly as before in a neighborhood a_3 of the point a and determines a solution $(x; y)$.

As an example consider the equation

$$(12). \qquad\qquad y - e \sin y = x \qquad\qquad (0 < e < 1),$$

which in the theory of elliptic orbits determines the value of the eccentric anomaly y in terms of the mean anomaly x. The function φ is in this case

$$\varphi(x, y, a, b) = \frac{e(\sin y - y \cos b) + x}{1 - e \cos b}$$

and φ_y remains less than θ when

$$| y - b | < \theta \frac{1 - e}{e} = \epsilon.$$

For any given $x = a$, a value $y = b$ can be determined, by graphical methods for example, so that

$$| \varphi(a, b, a, b) - b | = \left| \frac{b - e \sin b - a}{1 - e \cos b} \right| < \theta \frac{1 - e}{e} (1 - \theta).$$

The process described above therefore converges in a suitably chosen neighborhood of $x = a$, and a solution of equation (12) can be found when an approximate solution only has been determined in advance.

§ 4. Bolza's Extension of the Fundamental Theorem*

The neighborhood P_ϵ of a set of points P in the space $(x; y)$ is the totality of points $(x; y)$ which satisfy inequalities of the form

$$| x - a | < \epsilon, \qquad | y - b | < \epsilon,$$

---

* Vorlesungen über Variationsrechnung, page 160: also *Mathematische Annalen*, vol. 63 (1906), page 247. The theorem was proved independently by Mason and Bliss, "Fields of extremals in space," *Transactions of the American Mathematical Society*, vol. 11 (1910), page 326.

where $(a; b)$ is some point of P. The sets of points (a) and (b) which belong to points $(a; b)$ of P are the projections of P in the x- and y-spaces, and will be denoted by A and B, respectively.

The fundamental theorem of § 1 remains true if in its statement the single point p is replaced by a set of points P which is finite and closed, and which furthermore has the property that no two distinct points $(a; b)$, $(a'; b')$ of P have the same projection $a' = a$. According to the conclusions of the theorem there exists then a neighborhood P_ϵ in which no two solutions of the equations $f(x; y) = 0$ have the same projection x, and a neighborhood A_δ in which every x surely belongs to a solution $(x; y)$ in P_ϵ. The single-valued functions $y(x_1, x_2, \cdots, x_m)$ so defined in A_δ are continuous, and if the functions $f(x; y)$ have continuous derivatives of the n-th order in a neighborhood of P, so have the functions $y(x_1, x_2, \cdots, x_m)$ in A_δ.

To prove the theorem suppose first that a sequence of positive constants ϵ_k ($k = 1, 2, \cdots$) approaching zero has been selected arbitrarily. If the first part of the theorem were not true, then in any neighborhood P_{ϵ_k} there would be two distinct solutions $(x; y)_k$ and $(x; y')_k$ of the equations $f(x; y) = 0$, which would satisfy, respectively, inequalities of the form

$$(13) \qquad \begin{array}{ll} |x - \alpha| < \epsilon_k, & |y - \beta| < \epsilon_k; \\ |x - \alpha'| < \epsilon_k, & |y' - \beta'| < \epsilon_k \end{array}$$

with two points $(\alpha; \beta)_k$ and $(\alpha'; \beta')_k$ of the set P. Since P is finite and closed, the sequence of values $(\alpha, \beta; \alpha', \beta')_k$ has a point of condensation $(a, b; a', b')$ for which $(a; b)$ and $(a'; b')$ are both in P. From the inequalities (13) it follows that $(a, b; a', b')$ is also a point of condensation for the sequence $(x, y; x, y')_k$, and therefore a and a' must be the same. The values b and b' must also be identical since P contains only one point $p(a; b)$ with the projection a. According to the original statement of the fundamental theorem in § 1, a neighborhood p_ϵ can be chosen in which no two solutions of the equations $(x; y) = 0$ have the same projection x. Hence the existence of the sequences $(x; y)_k$ and $(x; y')_k$ with the common point of condensation $(a; b)$ is contradicted, and it must always be possible to select a neighborhood P_ϵ in which distinct solutions of the equations $f = 0$ always have distinct projections x.

A similar argument shows that a neighborhood A_δ can be selected so that to any point of it there corresponds a solution of the equations $f = 0$. Otherwise to each δ_k of a sequence of constants approaching zero, there would correspond a point $(x)_k$ in the region A_{δ_k} which would belong to no solution in P_ϵ. To each $(x)_k$ there would correspond a point $(\alpha)_k$ in A satisfying the inequalities

$$| x - a | < \delta_k$$

with the values $(x)_k$, and the points $(\alpha)_k$ would have a point of condensation a in A, which would also be a point of condensation for the sequence $(x)_k$, since A is finite and closed when P is so. But by the original theorem of § 1, again, it is known that a neighborhood a_δ of a can be chosen in which every point x has associated with it a solution $(x; y)$ in p_ϵ, where $p(a; b)$ is the point of P having the projection a. Consequently the existence of the sequence $(x)_k$ is contradicted.

If now the region P_ϵ is so restricted that the functional determinant $D(x; y)$ remains different from zero throughout it, then the original theorem of § 1 can be applied to show that the functions $y(x_1, x_2, \cdots, x_n)$ are continuous at any point of the region A_δ and possess as many continuous derivatives as are possessed by the functions $f(x; y)$.

§ 5. The Unique Sheet of Solutions Associated with an Initial Solution

The points of the space $(x; y)$ may be divided into two classes, ordinary points and exceptional points, with respect to the functions f. An *ordinary point* is one at which the first and third hypotheses of the theorem of § 1 are postulated, that is, one near which the functions f and their first derivatives with respect to y are continuous and the functional determinant $D = \partial(f_1, f_2, \cdots,$

$f_n)/\partial(y_1, y_2, \cdots, y_n)$ is different from zero. An *exceptional point* is one at which some of these conditions are not fulfilled or are not presupposed.

A *sheet of points* in the $(m + n)$-dimensional space $(x; y)$ may be defined as a point set S with the property that for any point $p(a; b)$ belonging to the set a neighborhood p_ϵ can always be found such that no two points of S in p_ϵ have the same projection x. In other words, the variables y are single-valued functions $y(x_1, x_2, \cdots, x_m)$ in the neighborhood of the point p, for points of the sheet.

If for any neighborhood b_ϵ of the kind just described, a region a_δ ($\delta \leq \epsilon$) can be found in which every point x belongs to a point of S in p_ϵ, then p is said to be an *interior point* of the sheet S.

A *boundary point* is a limit point of points of the sheet, which is not itself an interior point and may not even belong to S.

A sheet is said to be *connected* if every pair $(x'; y')$, $(x''; y'')$ of its interior points can be joined by a continuous curve

$$x = x(t), \qquad y = y(t) \qquad (t' \leq t \leq t''),$$

consisting entirely of interior points of the sheet.

In the following pages it is always to be understood that the sheets considered are continuous and have continuous first derivatives, or in other words at any interior point of one of them the functions $y(x_1, x_2, \cdots, x_m)$ mentioned above have these properties. A sheet will be said to become infinite near a point x' if x' is the limit of the projections of a sequence of points $(x; y)$ of the sheet for which one at least of the variables y approaches infinity.

With the preceding agreements as to nomenclature in mind, it is possible to prove the following theorem:

If a point $p(a; b)$ is an ordinary point for the functions f and satisfies the equations $f = 0$, then there passes through p one and only one connected sheet of solutions of these equations, with the properties:

1) *all points of the sheet are ordinary points of the functions f;*

2) *all points are interior points;*

3) *the only boundary points of the sheet are exceptional points for the system f.*

The set of points

$$[x_1, x_2, \cdots, x_m;\ y_1(x_1, x_2, \cdots, x_m),\ \cdots,\ y_n(x_1, x_2, \cdots, x_m)]$$

defined over the region a_s by the principal theorem of §1, is a sheet S_1 of solutions of the equations $f = 0$ which satisfies all the requirements of the theorem just stated except possibly the last. Its points are all interior points since the region a_s is defined by inequalities only. If any boundary point $p'(a'; b')$ of S_1 is an ordinary point of the functions f it must satisfy the equations $f = 0$, since the f's are continuous and p' is a limit point of points on S_1. Consequently the theorem of §1 can be applied in the neighborhood of p', and the sheet S' so determined near p' forms with S_1 a new set S_2. This process may be repeated any number of times, and the totality of points which can be attained by a finite number of such extensions, constitutes the sheet S required in the theorem.

The set of points S so determined constitutes a sheet, since any point q of it is an ordinary point and a solution of the equations $f = 0$, and according to the theorem of §1 the solutions of these equations in the neighborhood of q have the property which is characteristic of a sheet. From the manner of its construction the sheet is evidently connected and consists entirely of interior points. If any boundary point q of S were an ordinary point of the functions f, the sheet could be extended to include q as an interior point by the process described in the preceding paragraph.

There could not be a second sheet Σ containing a point π not in S and having the properties stated in the theorem. For there would in that case be a continuous curve

$$x = x(t), \quad y = y(t) \qquad (t_1 \leqq t \leqq t_2)$$

in Σ joining p with π and consisting entirely of ordinary points. In a neighborhood of $t = t_1$ all of the points defined on the curve would also be points of S, since the solutions of the equations

$f = 0$ near the initial point p of the curve are all in S. The values of t defining points on the curve and in S would therefore have an upper bound $\tau \leqq t_2$ such that τ would define on the curve a boundary point of S. But this is impossible since all of the points of the curve are ordinary points.

If the functions f are known to be continuous and to have continuous derivatives in a region R, then it follows readily from what precedes that through any ordinary solution of the equations $f = 0$ interior to R there passes one and only one sheet of solutions having the property that the only boundary points of the sheet are boundary points of R, or interior points of R at which the functional determinant vanishes. If R is finite and closed and consists entirely of ordinary points for the functions f, then there can not be more than a finite number of points of the sheet on any ordinate x. Otherwise the points common to the ordinate and the sheet would have a point of condensation p, also in R. Since p is an ordinary point there can be at most one solution of the equations in a properly chosen neighborhood p_ϵ.

It is interesting to determine a criterion which shall characterize a sheet which is at most single-valued on any ordinate. Such a criterion is derived in § 7 in connection with a theorem due originally to Schoenflies, and afterwards proved by Osgood. The proof of it involves the auxiliary notions described in § 6 and the following corollaries to the preceding theorem:

If the initial point of a continuous arc

$$(C_x) \qquad x_i = x_i(t) \qquad (i = 1, 2, \cdots, m; t' < t < t'')$$

in the x-space is the projection of a solution $p'(x'; y')$ of the equations $f = 0$ which is an ordinary point for the functions f, then there is associated with the arc C_x one and only one continuous curve

$$(C_{xy}) \quad x_i = x_i(t), \quad y_\alpha = y_\alpha(t) \qquad (i = 1, 2, \cdots, m; \alpha = 1, 2, \cdots, n)$$

passing through $(x'; y')$ for $t = t'$, with the properties:
1) *all of its points are solutions of the equations $f = 0$ and ordinary points of the functions f;*

2) *it is defined either over the whole interval $t' \leqq t \leqq t''$, or else on an interval $t' \leqq t < \tau \, (\leqq t'')$ such that as t approaches τ the only limit points of the curve C_{xy} are at infinity or are exceptional points of the functions f.*

The truth of this statement is readily deduced from the considerations which precede, or by the following argument. The fundamental theorem of § 1 can be applied at the point $(x'; y')$. If the arc C_x is entirely within the region x_δ' then the existence and uniqueness of the curve C_{xy} is evident. In any case there will be some intervals $t' \leqq t \leqq t_1$ in which curves C_{xy} are defined having all the properties described in the theorem except possibly 2). Suppose that τ is the upper bound of the end values t_1 for such intervals. Then there is a curve C_{xy} well defined in the interval $t' \leqq t < \tau$, and no limit point $(\alpha \, ; \beta)$ of the curve as t approaches τ can be a finite ordinary point for the functions f. For if $(\alpha \, ; \beta)$ were such a point, it would also satisfy the equations $f = 0$, on account of the continuity of the functions f, and the theorem of § 1 could again be applied at $(\alpha \, ; \beta)$. A curve C_{xy} with all the properties of the theorem, except possibly 2), could then be defined over an interval including the interval $t' \leqq t < \tau$ in its interior, which contradicts the assumption that τ is the upper bound of such intervals.

There could not be two curves C_{xy} associated with the projection C_x, having the properties described in the theorem, and having distinct points $(x; y')$ and $(x; y'')$ corresponding to the same value t_2. For if so, there would be an interval $t_3 < t \leqq t_2$ in which the curves would be distinct while at $t = t_3$ they coincide. This is, however, impossible since in a neighborhood of the point corresponding to t_3 there can be but one solution of the equations $f = 0$ corresponding to a given set of values x.

Suppose that a continuum X of points $(x_1, x_2, \cdots, x_m)$ contains no projection of a boundary point of a sheet S of solutions of the equations $f = 0$, and no point near which the sheet becomes infinite. Then if X contains the projection of a point on the sheet every other point of X will also be such a projection. On the other hand, if X

contains a point which is not a projection of any point of the sheet, then no point of X can be a projection of a point of S.

These statements follow readily with the help of the last theorem. For suppose that X contains the projection x' of an interior point $(x'; y')$ of a sheet of solutions of the equations $f = 0$, and let x'' be any other point of X. Since X is a continuum there exists a continuous arc C_x entirely interior to X joining x' and x'', and the corresponding continuation curve C_{xy} must be defined over the whole of the arc C_x. Hence x'' is also the projection of a point of the sheet of solutions through $(x'; y')$. The rest of the theorem follows at once.

If the curve C_{xy} in the last theorem but one is defined over the whole arc C_x, and has initial and end points p' and p'', respectively, then there always exists a positive constant ρ such that any curve Γ_x lying in the ρ-neighborhood of the curve C_x and joining x' to x'', has a unique continuation curve Γ_{xy} also joining p' and p''.

The curve

$$(\Gamma_x) \qquad x = \xi_i(u) \qquad (i = 1, 2, \cdots, m;\ u' \leqq u \leqq u'')$$

is said to lie in the ρ-neighborhood of C_x if there exists a continuous function

$$(14) \qquad\qquad t = t(u) \qquad\qquad (u' \leqq u \leqq u'')$$

taking the values t', t'' at the ends of the u-interval, and such that the point α on Γ_x, defined by any value of u, lies in the neighborhood a_ρ of the corresponding point a of C_x determined by the relation (14).

It is possible to choose two constants, ϵ and $\delta \leqq \epsilon$, so that the neighborhoods p_ϵ and a_δ have the properties described in the theorem of § 1 uniformly for every point $p(a, b)$ on the arc C_{xy}. If not, there would be a sequence of points p_k on C_{xy} with a limit point π, for which the largest possible constants ϵ_k have the limit zero. But for the point π there is an effective constant $\epsilon > 0$, and the constants ϵ_k could not therefore decrease indefinitely in size. A similar argument shows the existence of the constant δ.

Suppose now that the interval $u' \leqq u \leqq u''$ is divided by values u_k ($k = 1, 2, \cdots, \nu$) into sub-intervals so small that the points of any arc $\alpha_{k-1}\alpha_k$, corresponding on Γ_x to the values $u_{k-1} \leqq u \leqq u_k$, all lie in the $\frac{1}{2}\delta$-neighborhood of the point α_{k-1}, and further so small that the same is true with respect to the point a_{k-1} of the arc $a_{k-1}a_k$ of C_x corresponding to $\alpha_{k-1}\alpha_k$ by means of the relation (14). The constant ρ is supposed to have

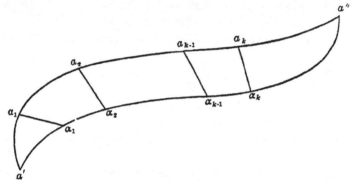

Fig. 1.

been chosen equal to $\frac{1}{2}\delta$, so that the curve Γ lies in the $\frac{1}{2}\delta$-neighborhood of C. Then the four-sided closed curve formed by the two straight lines $a_{k-1}\alpha_{k-1}$ and $a_k\alpha_k$, and the two arcs $a_{k-1}a_k$ and $\alpha_{k-1}\alpha_k$, lies entirely within the δ-neighborhood of the point a_{k-1}. The two continuation curves in the xy-space, starting with the point p_{k-1} on C_{xy} and having as projections the arcs $a_{k-1}a_k\alpha_k$ and $a_{k-1}\alpha_{k-1}\alpha_k$, respectively, lead to the same point π_k corresponding to the point α_k in the x-space.

It is possible to argue, then, that the point π_1 on the continuation curve of the arc $a'\alpha_1$ is the same as that of the continuation curve for $a'a_1\alpha_1$, since the arcs $a'\alpha_1$ and $a'a_1\alpha_1$ lie entirely within the δ-neighborhood of the point a_1. Similarly, the point π_2 for the arc $a'\alpha_2$ is the same as that for the continuation curve along $a'a_2\alpha_2$. And finally the point π'' must coincide with p'', provided always that the initial points π' and p' of the continuation curves are the same.

In particular if the curve C_{xy} is defined over the whole arc C_x, as described above, then there exists a polygon in the x-space joining a' and a'' in the ρ-neighborhood of C_x, and along which there is a continuation curve in S also joining p' and p''. The polygon can be so chosen that no two adjacent sides have more than an end point in common.

To show this, let the interval $t' \leqq t \leqq t''$ be divided in any way by means of points of division t', t_2, t_3, $\cdots$, t_ν, t'', and let the corresponding points on the curve C_{xy} be $(x'; y')$, $(\xi''; \eta'')$, $\cdots$, $(\xi^{(\nu)}; \eta^{(\nu)})$, $(x''; y'')$. The straight line $\xi^{(k)}\xi^{(k+1)}$ has the equations

$$x_i = \xi_i^{(k)} + \frac{t - t_k}{t_{k+1} - t_k}(\xi_i^{(k+1)} - \xi_i^{(k)}) \quad (i = 1, 2, \cdots, m).$$

Since the functions defining C_x are continuous, and therefore uniformly continuous, in $t' \leqq t \leqq t''$, it is possible to take the points of division t', t_2, t_3, $\cdots$, t_ν, t'' so close together that the differences $x - \xi^{(k)}$, for any point x on the arc $\xi^{(k)}\xi^{(k+1)}$ of C_x, are uniformly less than an arbitrarily assigned positive constant δ; and the preceding theorem shows that the curve C_{xy} and the continuation curve along the polygon both lead from p' to p''.

If the sides $\xi^{(k)}\xi^{(k+1)}$ and $\xi^{(k+1)}\xi^{(k+2)}$ have more than the point $\xi^{(k+1)}$ in common, then one of the two would be included entirely within the other, and the continuation curve along $\xi^{(k)}\xi^{(k+2)}$ would have the same end points as that along the two successive sides. Therefore, by replacing adjacent sides by a single one whenever the two have more than one end point in common, a polygon as described in the theorem can be found.

§ 6. AUXILIARY THEOREMS AND DEFINITIONS

In this section it is proposed to record some theorems which will be of service later, especially in the proofs of the theorems of § 7. In the first place let it be agreed that a regular curve in the plane shall mean one which is continuous and has a well-defined tangent at all except possibly a finite number of points, at each of which, however, the slope of the tangent approaches definite limits as the point is approached from either side. Analytically this means that the functions

$$x = x(t), \qquad y = y(t) \qquad\qquad (t' \leqq t \leqq t'')$$

defining a regular curve are continuous in the whole interval $t' \leqq t \leqq t''$, that they are differentiable and satisfy the inequality

$$(15) \qquad\qquad (dx/dt)^2 + (dy/dt)^2 \neq 0$$

at all except possibly a finite number of values of t. At an exceptional value $t = \tau$, where the derivatives are not well defined or where the expression (15) vanishes, the angle φ defined by the equations

$$\cos \varphi = \frac{dx/dt}{\sqrt{(dx/dt)^2 + (dy/dt)^2}}, \quad \sin \varphi = \frac{dy/dt}{\sqrt{(dx/dt)^2 + (dy/dt)^2}}$$

has nevertheless a unique limit as t approaches τ on the right, and a unique limit as t approaches τ on the left. These two limits are not necessarily the same.

It is known that a simply closed regular curve C in an xy-plane divides the plane into two continua, an exterior and a finite interior.* Any two interior points can be joined by a regular curve every point of which is an interior point, and a similar statement holds for exterior points. Any continuous curve joining an interior and an exterior point must have on it at least one point of the curve C, and any point p on C can be joined with an interior point by a regular curve which has in common with C only the point p.

The interior of a simply closed regular curve

$$x = x(t), \qquad y = y(t) \qquad\qquad (t' \leqq t \leqq t'')$$

can be divided by a finite number of segments of straight lines into

---

* See for example Osgood, Lehrbuch der Funktionentheorie, Chapter V, §§ 4–6; Bliss, "A proof of the fundamental theorem of analysis situs," *Bulletin of the American Mathematical Society*, vol. 12 (1906), page 336.

regions each of which has a maximum diameter less than an arbitrarily assigned positive constant ϵ.

Let the maximum and minimum values of y in the interval $t' \leqq t \leqq t''$ be y_1 and y_2, and let p_1 and p_2 be two points of C at which y has these values. It is desired to show that there is a segment $p'p''$ of the horizontal line $y = (y_1 + y_2)/2$ which forms with C two simply closed regular curves, $p'p_1p''p'$ and $p'p_2p''p'$, each containing one of the points p_1 and p_2.

The points p_1 and p_2 can be joined by a regular curve D which, except at its end points, is interior to C. Two arcs of D adjoining p_1 and p_2, can be marked off in such a way that they do not cut the line $y = (y_1 + y_2)/2$. The remaining arc D' of D is entirely interior to C and can be replaced by a continuous polygon D'' with a finite number of sides, having the same end points and consisting also of interior points of C only. Any side of D'' which has an end point in common with the line $y = (y_1 + y_2)/2$ may be rotated slightly about its other end point, and in this way it may be brought about that D'' has only interior points of its sides on the line $y = (y_1 + y_2)/2$, and actually crosses the line wherever they have a point in common.

The polygon D'' must intersect $y = (y_1 + y_2)/2$ at least once, say at a point p, since one end point of D'' is above and the other below this line. There will be a segment $p'p''$ of $y = (y_1+y_2)/2$, containing p and such that p' and p'' are on the curve C while every other point of the segment is interior to C. There can be only a finite number of such segments $p'p''$ containing points of D'', since D'' has at most a finite number of intersections with the horizontal line. There must be at least one segment on which D'' has an odd number of intersection points, since otherwise both end points of D'' would be on the same side of $y = (y_1 + y_2)/2$. If $p'p''$ is such a segment, then it forms with C two simply closed regular curves $p'p_1p''p'$ and $p'p_2p''p'$, one of which contains p_1 and the other p_2. For after its last intersection with $p'p''$ the polygon D'' and hence p_2 is entirely exterior to the curve $p'p_1p''p'$.

---

* For a similar theorem see Osgood, loc. cit., Chapter V, § 9.

For the moment that part of a curve which does not lie in a horizontal line may be called the effective arc of the curve, in view of the fact that the altitude of the curve can not be more than one half the length of this so-called effective part. If the altitude of any curve is $\geq \epsilon$, the effective length of either of its two parts after subdivision by a horizontal line segment, as described above, will be $\leq L - \epsilon$, where L is its effective length.

If the altitude $y_1 - y_2$ of C is greater than ϵ, then the effective arc of either $p'p_1p''p'$ or $p'p_2p''p'$ will be greater in length than ϵ, and the effective length of each will also be less than $L - \epsilon$, where L is the perimeter of C. If the curve $p'p_1p''p'$, for example, has still an altitude greater than ϵ, it may be subdivided by a horizontal segment as before, and the effective arcs of the two new curves so found will be less than $L - 2\epsilon$. By a continuation of this process the interior of C will be subdivided finally by curves whose effective lengths are less than 2ϵ and whose altitudes are therefore less than ϵ.

In a similar manner the regions so formed may be subdivided by vertical segments into others whose breadths are less than ϵ, and the theorem follows at once.

A set of points in an x_1x_2-plane is *connected* if any two of its points can be joined by a continuous arc whose points all belong to the set, and it is further said to be *simply connected* if every simply closed regular curve in it has an interior which also consists only of points of the set.

It is more difficult to set down a satisfactory definition of simple connectivity for sets of points in an m-dimensional space. In the following section of these lectures, however, a special type of simple connectivity is needed which may be defined by means of some simple auxiliary conceptions.

A *normal subspace of two dimensions* in a region X of points $(x_1, x_2, \cdots, x_m)$ is a totality of points defined by equations of the form

$$x_i = \varphi_i(u_1, u_2) \qquad (i = 1, 2, \cdots, m),$$

where

1) the values (u_1, u_2) range over a simply connected region U;

2) no two distinct sets of values u define the same point x;

3) the functions φ are continuous and have continuous first derivatives in U;

4) the determinants of the second order of the matrix of derivatives $\|\partial\varphi_i/\partial u_k\|$ $(i = 1, 2, \cdots, m; k = 1, 2)$ do not all vanish simultaneously at any point of U.

A simply connected region in two dimensions is defined above, and a connected region X in a space of points $(x_1, x_2, \cdots, x_m)$ has a definition quite similar to that for two dimensions. In order to specify conveniently the properties of a region X which is simply connected, the term *elementary curve* will also be used. By an elementary curve in X is meant a simply closed continuous curve which either lies in a normal subspace of two dimensions entirely in the interior of X, or else is such that in every neighborhood of it there is a simply closed continuous curve having this property. It is thus seen that while an elementary curve may not itself be imbedded in one of the two-dimensional normal subspaces interior to X, it can nevertheless be approximated as closely as may be desired by one which does. The word neighborhood is here used in the sense described in connection with the fourth theorem of § 5 (see page 26).

If a region X is connected, then any simply closed continuous curve in its interior may be developed into two such curves by an auxiliary arc joining two of its points, and the process of development may be continued on the two arcs so formed.

If a region X is such that any simply closed continuous curve in its interior is an elementary curve, or may be developed into a number of elementary curves by means of auxiliary arcs, as just described, then X is said to be simply connected.*

· * For a discussion of the connectivity of higher spaces, see Picard and Simart, Théorie des Fonctions algébriques de deux Variables indépendantes, Chapître II, in particular §§ 11 ff. If every simply closed continuous curve interior to R lies in a normal subspace of two dimensions interior to R, one sees intuitively that a second neighboring subspace of the same kind can be passed through the curve. The closed two-dimensional subspace so formed is

§ 7. A Criterion that a Sheet of Solutions be Single-Valued

Consider in the first place a set of equations

$$(16) \qquad f_a(x_1, x_2; y_1, y_2, \cdots, y_n) = 0 \qquad (\alpha = 1, 2, \cdots, n)$$

in which there are but two independent variables x.

If a connected sheet S of solutions of equations (16) *consists only of ordinary points of the functions f, and furthermore has a simply connected projection X in the x_1x_2-plane such that no interior point of X is either a point where S becomes infinite or the projection of a boundary point of S, then the sheet S is single-valued over the interior of X.*

Suppose, in contradiction to the theorem, that over any interior point of X there were two points, p' and p'', of the sheet. Since S is connected there would be a continuous curve

$$(C_{xy}) \quad x_1 = x_1(t), \quad x_2 = x_2(t), \quad y_a = y_a(t)$$
$$(t' \leqq t \leqq t''; \; \alpha = 1, 2, \cdots, n)$$

consisting entirely of interior points of the sheet and joining p' with p'' in the space $(x; y)$. The projection

$$(C_x) \qquad\qquad x_1 = x_1(t), \qquad x_2 = x_2(t) \qquad (t' \leqq t \leqq t'')$$

of this curve would necessarily be a closed curve in the x_1x_2-plane, and by the second theorem of § 5 the arc C_{xy} is the only one associated with C_x in the sheet S and having the initial point p'.

The curve C_x may be simply closed and regular; but if it is not, there will nevertheless be a curve in the region X having these properties, and for which the continuation curve analogous to C_{xy} is not closed. For, in the first place, from § 5 it is seen that the curve C_x may be supposed to be a polygon no two adjacent sides of which have more than an end point in common, provided that it is desired only to secure a continuous curve in

---

separated into two parts by the curve, and hence the number which Picard and Simart designate by p_1 is equal to unity for a simply connected region of the kind defined in the text above.

4

the sheet passing from p' to p''. Let the corners of this polygon in the x-plane be denoted by $\xi_1, \xi_2, \cdots, \xi_\nu$, where ξ is a symbol for a point (x_1, x_2). The side $\xi_\nu \xi_1$ touches $\xi_1 \xi_2$ at its end point ξ_1, and it can be argued therefore that there will be some first side $\xi_\lambda \xi_{\lambda+1}$ which touches some one of the preceding sides elsewhere than at its initial point ξ_λ. Let the side so touched by $\xi_\lambda \xi_{\lambda+1}$ be $\xi_\kappa \xi_{\kappa+1}$, where $\kappa + 1$ is necessarily less than λ, and let the first point of $\xi_\lambda \xi_{\lambda+1}$ which lies on $\xi_\kappa \xi_{\kappa+1}$ be ξ. If the portion of the curve C_{xy} which corresponds to the polygon

$$(17) \qquad \xi, \quad \xi_{\kappa+1}, \quad \xi_{\kappa+2}, \quad \cdots, \quad \xi_\lambda, \quad \xi$$

is not closed, then the polygon (17) itself is a simply closed curve in X of the kind desired above, that is, one along which there exists a continuation curve in the xy-space whose end points are different.

If the portion of C_{xy} which corresponds to (17) is closed, then that part of C_{xy} which belongs to the polygon

$$(18) \qquad \xi_1, \quad \xi_2, \quad \cdots, \quad \xi_\kappa, \quad \xi, \quad \xi_{\lambda+1}, \quad \cdots, \quad \xi_\nu, \quad \xi_1$$

is also continuous and leads from p' to p''. Since $\kappa + 1 < \lambda$ the side $\xi_{\kappa+1} \xi_{\kappa+2}$ at least is missing in (18), and the number of sides is at least one less than that of the original polygon. By an alteration of the kind suggested in the proof of the last theorem of § 5, which also reduces the number of sides, it can be brought about, if not already true, that the polygon (18) still has no two adjacent sides with more than an end point in common.

By continuing this process one must come at some stage to a simply closed regular curve in the x-plane with a corresponding continuation curve in the xy-space which is not closed. In order not to complicate the notation too much it may be supposed that the curve C_x itself is such a curve. Every point of C_x is an interior point of the region X since the corresponding point of C_{xy} is an interior point of the sheet S. The interior of C_x is therefore also composed entirely of interior points of X, since X is simply connected. If the interior of C_x is subdivided into

two parts by a segment of a straight line, as described in the preceding section, the dividing segment will also have a continuation curve on the sheet S throughout its entire length, by the
second theorem of § 5. For its initial point on the curve C_x
corresponds to an interior point of the sheet S and, by the hypothesis of the theorem which is to be proved, none of its points
can be a point where S becomes infinite or can correspond to a
boundary point of S. Hence one of the simply closed curves
formed by the curve C_x and the dividing segment is a curve
retaining the property that it has a continuation curve on
the sheet S which is not closed. Suppose that C_x' is this curve.
By continuing the process a sequence of curves $\{C_x^{(k)}\}$, with
diameters approaching zero, can be found, each lying in the
interior of C_x and having an unclosed continuation curve $C_{xy}^{(k)}$
on S.

If a point $p^{(k)}$ is selected arbitrarily on the curve $C_{xy}^{(k)}$, the
sequence $\{p^{(k)}\}$ $(k = 1, 2, \cdots, \infty)$ will have a finite point of condensation $\pi(\alpha; \beta)$ in the xy-space which is an interior point of the
sheet S. For the projections $x^{(k)}$ of the points $p^{(k)}$ all lie in the interior of C_x and hence must have a point of condensation α. Furthermore the points of the sequence $p^{(k)}$ whose projections are in the
neighborhood of α can not become infinite or approach a boundary
point of the sheet, since α is interior to X. They must therefore
have at least one limit point π which is an interior point of the
sheet, and with which there are associated two neighborhoods
π_ϵ and α_δ by the principal theorem of § 1. Some of the points
$p^{(k)}$ lie in π_ϵ, and have corresponding curves $C_x^{(k)}$ in α_δ. For
such points the continuation curves $C_{xy}^{(k)}$ also lie in π_ϵ and can
not be unclosed, since to any point x in α_δ there corresponds in
π_ϵ at most one solution of the equations $f = 0$. The original
assumption that S is multiple-valued in the interior of X is
therefore contradicted.

The theorem remains true for any system of equations of the form

$$(19) \quad f_a(x_1, x_2, \cdots, x_m; y_1, y_2, \cdots, y_n) = 0 \quad (\alpha = 1, 2, \cdots, n).$$

In this case the curves C_{xy} and C_x have equations

(C_{xy})
$$x_i = x_i(t), \qquad y_\alpha = y_\alpha(t)$$
$$(i = 1, 2, \cdots, m; \alpha = 1, 2, \cdots, n; t' \leqq t \leqq t''),$$

(C_x)
$$x_i = x_i(t),$$

and the question asked in the proof of the theorem just stated is whether or not the latter curve may be closed while the former has distinct end points.

It is a part of the hypothesis of the theorem that the region X is simply connected according to the definition of the preceding section; and, according to the arguments made in the paragraphs above, the curve C_x may be supposed a simply closed polygon. In any neighborhood of C_x there will be, according to § 6, on account of the simple connectivity, an elementary curve $\bar{C}_x$ lying in a normal subspace of two dimensions

$$(20) \qquad\qquad x_i = g_i(u_1, u_2) \qquad\qquad (i = 1, 2, \cdots, m)$$

entirely interior to X. If the continuation curve C_{xy} is not closed, and if $\bar{C}_x$ is taken sufficiently near to C_x, then the corresponding continuation curve $\bar{C}_{xy}$ will also not be closed.

The normal subspace (20) is defined over a simply connected domain U of points (u_1, u_2), and has no multiple points. To every point of $\bar{C}_x$ there corresponds therefore a single pair of values

(C_u)
$$u_1 = u_1(t), \qquad u_2 = u_2(t);$$

and the functions so defined are continuous, by the principal theorem of § 1, since at every point some pair of the equations (20) has a functional determinant for u_1, u_2 which is different from zero. The curve corresponding to the curve $\bar{C}_{xy}$ in the space $(u; y)$ may be denoted by

(C_{uy})
$$u_1 = u_1(t), \quad u_2 = u_2(t), \quad y_\alpha = y_\alpha(t), \quad (\alpha = 1, 2, \cdots, n),$$

and its initial point, corresponding to p', by p_u' $(u'; y')$. Every point of C_{uy} is an ordinary solution of the equations

$$(21) \quad \varphi_\alpha(u_1, u_2; y_1, y_2, \cdots, y_n) = f_\alpha(g_1, g_2, \cdots, g_m; y_1, y_2, \cdots, y_n) \parallel 0$$
$$(\alpha = 1, 2, \cdots, n).$$

With a continuous curve C joining (u_1', u_2') to an arbitrarily chosen point (u_1, u_2) of U there is always associated a continuation curve of solutions of the equations (21), having the initial point p_u' and defined throughout the whole of C, since any such curve defines a curve in the x-space interior to X along the whole of which there is a corresponding continuation curve for the equations (19) in the sheet S. Hence there is a unique sheet S_u of solutions of the equations (21) whose projection in the u_1u_2-space is U; and no interior point of U is a point where the sheet becomes infinite or corresponds to a boundary point of the sheet, since the same is true of S with respect to X. The preceding argument can therefore be applied to show that the sheet S_u is single-valued over the region U, and the existence of the curve C_{uv} with the distinct end points p_u' and p_u'' is contradicted. Hence C_{xy} can not have distinct end points p' and p'', and the theorem last stated is proved.

§ 8. Transformations of n Variables and a Modification of a Theorem of Schoenflies

It is interesting to deduce by means of the preceding theorems some conclusions concerning a system of equations of the form

$$(22) \quad f_a(x; y) = x_a - \psi_a(y_1, y_2, \cdots, y_n) = 0 \quad (\alpha = 1, 2, \cdots, n).$$

The functions ψ are once for all assumed to be single-valued, continuous, and to have continuous first derivatives in a continuum Y in which the functional determinant

$$D = \partial(\psi_1, \psi_2, \cdots, \psi_n)/\partial(y_1, y_2, \cdots, y_n)$$

is different from zero. By a continuum is meant a set of points consisting only of interior points any two of which can be connected by a continuous curve lying entirely within the set. The boundary points of Y will be denoted by B, and X will represent the set of points in the x-space which corresponds to Y by means of the equations (22).

Any sequence $\{y^{(k)}\}$ of points $(y_1^{(k)}, \; y_2^{(k)}, \; \cdots, \; y_n^{(k)})$ $(k = 1, 2, \cdots)$ in Y, which approaches infinity or has a point of B as limit point, defines a corresponding sequence of points $\{x^{(k)}\}$ in X. The set of points of condensation for such sequences $\{x^{(k)}\}$ will be denoted by A.

The totality of solutions of the equations (22) *corresponding to points of the continuum Y form a single connected sheet S whose only boundary points have projections x and y in the sets A and B, respectively.*

For suppose that $(x'; y')$ is a first solution and $(x''; y'')$ any other. The points y' and y'' can be joined by a continuous curve interior to Y

$$y_a = y_a(t) \qquad (\alpha = 1, 2, \cdots, n; \; t' \leqq t \leqq t''),$$

and the corresponding curve

$$x_a = x_a(t), \qquad y_a = y_a(t),$$

defined by equations (22), is a curve interior to the sheet S and joining $(x'; y')$ to $(x''; y'')$, so that S is evidently connected. Any boundary point $(\alpha; \beta)$ of S must be the limit of a sequence of points $p^{(k)}$ for which the projections y are in Y. The limit β of the sequence $y^{(k)}$ can not be in Y, since then $(\alpha; \beta)$, by the theorem of § 1, would be an interior point of S. Hence β must be in B and α in A.

One may say further that if $p^{(k)}$ is a sequence of points $(x^{(k)}; y^{(k)})$ in S for which the sequence $x^{(k)}$ approaches infinity, then the only finite points of condensation possible for the sequence $y^{(k)}$ are in B. The statement is true when x and y are interchanged, on account of the definition above of the set A.

If the points of the set A are distinct from those of the image X of Y, then X is a single continuum whose only boundary points are points of A.

To prove this, consider an arbitrarily chosen point y' of Y. None of the points in a suitably chosen neighborhood of the corresponding values x' are points of A, since by the fundamental theorem of § 1 all such points correspond by means of equations (22) to points of Y, and are therefore points of X. Consider now the continuum $\overline{X}$ consisting of all points x which can be joined to x' by continuous curves containing no points of A, a continuum to which the neighborhood of x' certainly belongs, as has just been shown.

All the points of X are in the continuum $\overline{X}$, since the solutions of equations (22) corresponding to points of Y form a single connected sheet S. The curve in S joining $(x'; y')$ with any other point $(x''; y'')$ of the sheet has therefore a projection in the x-space joining x' with x'' and containing no points of the set A.

All of the points of $\overline{X}$ are points of X. For any set of values x in $\overline{X}$ can be joined to x' by a continuous curve C_x lying entirely in $\overline{X}$ and containing therefore no points of A. By the second theorem of § 5 the corresponding continuation curve C_{xy} must extend along the entire arc C_x, since otherwise the value of y for points on C_{xy} would approach infinity or else have a limit point on the boundary B of F, and some point of C_x would in that case necessarily be a point of A. It follows that x, like x', is the image of some point y in Y.

From the initial theorem of the last section, for the case when there are more than two variables, it follows that

If A is distinct from X, and X is simply connected in the sense of §6, then the sheet S is single-valued. In other words the continuum Y is tranformed in a one-to-one way into a continuum X by means of the equations (22), and the functions

$$(23) \qquad y_\alpha = y_\alpha(x_1, x_2, \cdots, x_n) \qquad (\alpha = 1, 2, \cdots, n)$$

so defined over X are single-valued, continuous, and have continuous first derivatives.

The character of the functions (23) near any point of X follows at once from the theorem of § 1.

Let it be supposed that the set of points A divides the x-space into exactly two continua X, Ξ such that every point of A is a bound-

*ary point for each of them, and suppose furthermore that there is a
particular point ξ in Ξ which does not correspond by means of the
equations (22) to any point of Y. Then the image X of Y is
distinct from A and coincides with X. If X is simply connected
the other conclusions of the last theorem follow at once.*

In the first place it can be shown that if any point ξ' of Ξ
corresponds to a point of Y then every other point ξ'' of Ξ
would also have this property. For ξ' and ξ'' can be joined by
a continuous curve

$$x_a = x_a(t) \qquad (\alpha = 1, 2, \cdots, n; \; t' \leqq t \leqq t'')$$

entirely interior to Ξ. The corresponding continuation curve

$$x_a = x_a(t), \qquad y_a = y_a(t)$$

of solutions of equations (22) must be defined along the whole of
the interval $t' \leqq t \leqq t''$, since otherwise as t approached any
upper bound τ of the values t which could be reached by continuation, the corresponding points y of the curve would have
to approach infinity or else have a point of condensation on the
boundary of Y. But this is impossible, since for a sequence of
points x corresponding to a sequence of points in Y approaching
infinity or a boundary point of Y, the only limiting points possible
are at infinity or else in the set A. It follows at once, on account
of the hypothesis of the theorem, that no point of Ξ can correspond
to a point of Y, and neither can any point of A, since in any
neighborhood of such a point of A there are points of Ξ which
in that case would also correspond to values y in Y. The image
of the region Y in the x-space is a single continuum whose only
boundary points are points of A. According to the preceding
argument it cannot be Ξ and it must therefore be X.

A modification of a theorem of Schoenflies can be deduced
readily from the results which precede. The theorem has to do
with a pair of equations of the form

$$(24) \qquad x_1 = \psi_1(y_1, y_2), \qquad x_2 = \psi_2(y_1, y_2)$$

in which the functions ψ are single-valued, continuous, and have continuous derivatives on a simply closed regular curve B of the y-plane and in the interior Y of B. The functional determinant $D = \partial(\psi_1, \psi_2)/\partial(y_1, y_2)$ is supposed to be different from zero in Y.

If the curve A in the x-plane formed by transforming the simply closed regular curve B in the y-plane, by means of the equations (24), is distinct from the image X of the interior Y of B, then X is a simply connected continuum whose only boundary points are points of A, and the correspondence defined between X and Y is one-to-one. The single valued functions

$$(25) \qquad y_1 = y_1(x_1, x_2), \qquad y_2 = y_2(x_1, x_2),$$

so determined in the region X, are continuous and have continuous first derivatives.*

From the preceding theorems of this section it follows that the complete image X of Y is a single finite continuum whose only boundary points are points of A. It remains to show that X is simply connected and that the correspondence between X and Y is one-to-one.

If any simply closed regular curve C_x is drawn in X, its interior must consist entirely of points of X. Otherwise there would necessarily be a boundary point of X, a point of the curve A, interior to C_x, and there would also be points of A exterior to C_x since X is finite. Hence there would necessarily be a point of the continuous curve A on C_x itself, which contradicts the assumption that A and X are distinct. It follows at once from the first paragraphs of § 7 and the simple connectivity of X just proved, that only one point y in Y corresponds to a given x in X, and by the theorem of § 1 it may be seen that the functions

---

* Schoenflies assumed only the continuity of the functions ψ_1, ψ_2, adding, however, that the correspondence defined between the regions X and Y of the two planes is to be one-to-one. In the theorem here proved ψ_1 and ψ_2 are subjected to further continuity restrictions, but the correspondence is proved to be unique. See Schoenflies, "Ueber einen Satz der Analysis Situs," *Göttinger Nachrichten* (1899), page 282. The theorem was later proved by Osgood and Bernstein in the same journal (1900), pages 94 and 98, respectively.

(25) have the continuity properties described in the theorem in the neighborhood of any particular point x.

Another theorem, slightly different in form, may be stated as follows:

If the images of the points of the simply closed regular curve B in the y-plane all lie on a simply closed regular curve A in the x-plane, then the equations (24) define a one-to-one correspondence between the interior X of A and the interior Y of B, and the functions (25) so defined have the same continuity properties as before.

In this case it can first be shown that the image x' of any point y' in Y must be distinct from A, and the rest of the proof is the same as before. For, if x' were a point of A, every point of a properly chosen neighborhood of x' would also be the image of a point of Y, since at $(x'; y')$ the functional determinant of equations (24) does not vanish. It would follow then, by continuation, that every point exterior to the curve A would also be the image of a point of Y, which is impossible since the functions ψ are finite. The continuum X is therefore identical with the interior of A, by the preceding theorems, and the correspondence between X and Y is one-to-one.

An example applying some of the theorems of §§ 5, 8 is given at the end of § 14.

CHAPTER II

SINGULAR POINTS OF IMPLICIT FUNCTIONS

The theorems which have been developed in the preceding pages of these lectures have to do with the behavior of implicit functions at ordinary points, or in regions which have no singular points in their interiors. For singular points where the functional determinant vanishes the theory is much more complicated, and no methods which can be comprehensively applied have so far been developed. There are, however, many special cases in widely different fields which have been studied with success, and it may not be out of place to glance at a few of them before proceeding to the further theorems with which these pages are primarily concerned.

Perhaps the most complete single theory which has been developed is that which has to do with the singularities of an algebraic function y of x determined by an equation of the form

$$(1) \qquad P(x, y) = 0,$$

where P is an irreducible polynomial in the two variables x and y. Suppose for convenience that the singular point to be considered is at the origin, and that the polynomial $P(0, y)$ has a lowest term of degree n in y. Then it is known that for each value of x in a sufficiently small neighborhood of $x = 0$, there exist exactly n solutions y of equation (1) in the neighborhood of $y = 0$, and the values of these solutions are given by k cycles of the form

$$(2) \qquad y = a_j x^{\mu_j/p_j} + a_j' x^{\mu_j'/p_j} + \cdots \quad (j = 1, 2, \cdots, k),$$

where the numbers μ, p are positive integers satisfying the relations

$$\mu_j < \mu_j' < \mu_j'' < \cdots, \qquad p_1 + p_2 + \cdots + p_k = n.$$

The series is one member of the cycle; the others are found by replacing x^{1/p_j} by $\omega^{\nu}x^{1/p_j}$ ($\nu = 1, 2, \cdots, p_j - 1$), where ω is a primitive p_j-th root of unity. The number p_j has no factor in common with the exponents $\mu_j, \mu_j', \cdots$. Otherwise the expansion would be in terms of a root of x of lower order than p_j. Thus there are in all n series in fractional powers of x which define the roots of the algebraic equation in the neighborhood of the origin. The coefficients of the series may be computed by means of the well-known Newton polygon,* or by methods due to Hamburger[†] and Brill.[‡] If the substitution $x = t^{p_j}$ is made in the series (2), the points (x, y) which it defines may be expressed in the parametric representation

$$x = t^{p_j}, \qquad y = t^{\mu_j}\{\alpha_j + \alpha_j' t^{\mu_j' - \mu_j} + \cdots\} \qquad (j = 1, 2, \cdots, k).$$

All the solutions of the equation (1) in the neighborhood of the origin evidently belong to a finite number of such branches.

With the help of the preparation theorem of Weierstrass, which is to be studied in the following pages, results similar to those just given may be proved for the solutions of an equation $F(x, y) = 0$ in the vicinity of any point where F is analytic.

The singularities of a surface

$$F(x, y, z) = 0$$

at a point where the function F is analytic have also been extensively studied. The points of the surface in the neighborhood of a singular point are determined by means of a finite number of expansions of the form

$$x = P(u, v), \qquad y = Q(u, v),$$

where P and Q are analytic in the parameters u and v.§

---

* See Appell and Goursat, Théorie des Fonctions algébriques, pp. 184 ff.

† Weierstrass, Werke, vol. 4, Kapitel 1.

‡ Münchener Berichte, vol. 21 (1891), p. 207.

§ See Black, "The parametric representation of the neighborhood of a singular point of an analytic surface," Proceedings of the American Academy of Arts and Sciences, vol. 37 (1902), p. 281.

In the calculus of variations the construction of "fields of extremals" in the plane requires the study of the real solutions of a system of equations of the form

$$(3) \qquad x = \varphi(t, a), \qquad y = \psi(t, a).$$

The extremals are the curves in the xy-plane defined by these equations for different values of a. Suppose that the parametric values

$$(4) \qquad t_0 \leqq t \leqq t_1, \qquad a = a_0$$

define an arc E which does not intersect itself and which consists entirely of points where the functional determinant

$$(5) \qquad \Delta(t, a) = \frac{\partial(\varphi, \psi)}{\partial(t, a)}$$

is different from zero. Then to any point (x, y) in a properly chosen neighborhood of E there corresponds but one solution (t, a) of equations (3), in the neighborhood of the values (4); and the functions

$$t = t(x, y), \qquad a = a(x, y)$$

so defined have continuity properties similar to those of φ and ψ themselves.* The neighborhood thus simply covered by the extremals (3) is the "field," and is perhaps the simplest example of the notion since it consists only of non-singular solutions of the equations (3).

When it is desired to find an arc C which minimizes an integral with respect to variations lying entirely on one side of C, a field of a different sort can be constructed.† The equations of the

---

The mathematical literature concerned with the singularities of a curve or surface, particularly their transformation into simpler types, is very large. The reader is referred to Pascal, Repertorium der höheren Mathematik, 2d edition, vol. 2, erste Hälfte, pp. 291 ff; and Encyclopädie der Mathematischen Wissenschaften, II B 2, p. 119, and III C 4, pp. 365 ff.

* Bolza, Vorlesungen über Variationsrechnung, pp. 249 ff.

† Bliss, "Sufficient conditions for a minimum with respect to one-sided variations," *Transactions of the American Mathematical Society*, vol. 5 (1904), p. 477; Bolza, "Existence proof for a field of extremals tangent to a given curve," ibid., vol. 8 (1907), p. 399.

extremals (3) can be taken so that for $t = 0$ they all intersect C and are tangent to it, and the equations

$$x = \varphi(0, a), \qquad y = \psi(0, a)$$

will then be the equations of C. If the curvatures of the two arcs at their point of contact are always different, then the extremal arcs E simply cover a portion of the plane N on one side of C and adjacent to it. In other words, the equations (3) define a one-to-one correspondence between the points of a region adjoining the axis $t = 0$ in the ta-plane, shown in the accompanying figure, and a certain neighborhood N on one side of the arc C.

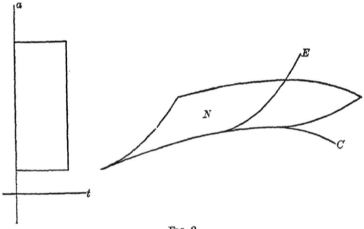

Fig. 2.

In the interior of the region N the functions $t(x, y)$, $a(x, y)$ have continuity properties similar to those of φ and ψ themselves. It is easy to see that this is a case in which the functional determinant (5) vanishes along the boundary $t = 0$ of the region to be transformed, since the curves C and E are always tangent.

In a paper published since these lectures were given, Dr. E. J. Miles* has considered the transformation defined by the equations

* " The absolute minimum of a definite integral in a special field," *Transactions of the American Mathematical Society*, vol. 13 (1912), pp. 37 ff.

(3) when the curve C to which the extremals E are tangent has a cusp, a situation corresponding to still another problem in the

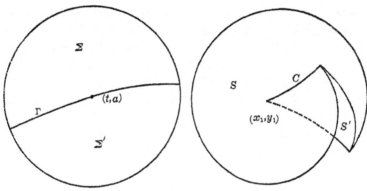

FIG. 3.

calculus of variations. In that case a point (t_1, a_1) and a curve Γ through it are transformed into a point (x_1, y_1) and a curve C as shown in the figure. One portion Σ of a neighborhood of (t_1, a_1) is then transformed in a one-to-one way into the leaf S, and the other portion Σ' into the leaf S'. At any point in the interior of one of the leaves, the variables t and a are single-valued functions of x, y having continuity properties similar to those of φ and ψ. The transformation is singular along the curve Γ.

The three examples which have been just described are only a few of the many proofs for the existence of fields involving transformations with singular points which might be cited.* Nearly all of these have to do with singularities of transformations of the form

(6) $$x = \varphi(u, v), \qquad y = \psi(u, v),$$

---

* Bliss, "The construction of a field of extremals about a given point," *Bulletin of the American Mathematical Society*, vol. 13 (1906), p. 47; Mason and Bliss, "Fields of extremals in space," *Transactions of the American Mathematical Society*, vol. 11 (1910), p. 325; Bill, "The construction of a space field of extremals," *Bulletin of the American Mathematical Society*, vol. 15 (1908), p. 374; Szücs, "Sur l'extrémale qui joint deux points donnés," *Mathematische Annalen*, vol. 71 (1912), p. 380. The method used by Szücs is quite closely that of Mason and Bliss in the paper mentioned above.

or

$$x = \varphi(u, v, w), \qquad y = \psi(u, v, w), \qquad z = \chi(u, v, w),$$

which have been studied also in a series of papers of more recent date presented as dissertations for the degree of doctor of philosopny at Harvard University.* The methods which have been used in the different cases have differed widely, and it does not seem possible at present to formulate a theory which includes them all. It is the intention of the writer, however, to show in the following pages how the transformation theorems proved above in § 7 may be applied to throw much light on the nature of real transformations of the form (6) in the neighborhoods of singular points. In the section of the lectures immediately following this introduction a simple algebraic proof of the preparation theorem of Weierstrass is given, not depending upon the theory of functions of a complex variable. A generalization of it is given in a later section which, in what might be called the general case, enables one to describe the behavior of the solutions of a system of equations of the form

$$f_i(x_1, x_2, \cdots, x_m; y_1, y_2, \cdots, y_n) = 0 \quad (i = 1, 2, \cdots, n)$$

in the neighborhood of a point where the functional determinant

$$\frac{\partial(f_1, f_2, \cdots, f_n)}{\partial(y_1, y_2, \cdots, y_n)}$$

vanishes. For these equations the variables x and y are permitted to have complex values.†

---

* Urner, " Certain singularities of point transformations in space of three dimensions," *Transactions of the American Mathematical Society*, vol. 13 (1912), p. 233; Clements, " Implicit functions defined by equations with vanishing jacobian," to appear in the same journal. Dederick, in a paper entitled " The solutions of an equation in two real variables at a point where both the partial derivatives vanish," *Bulletin of the American Mathematical Society*, vol. 16 (1909), p. 174, has discussed the singularities of a curve of the form $F(x, y) = 0$ with the help of a sort of generalization of the Weierstrass preparation theorem for a function which is not necessarily analytic.

† The proof given in these pages for the last-mentioned theorem is for the case of two variables y. For n variables see the reference in the last footnote to § 13.

§ 9. The Preparation Theorem of Weierstrass

The theorem which is to be proved may be stated in the following form:

Let $f(x_1, x_2, \cdots, x_m, y)$ be a convergent series in the variables x, y, and such that the series $f(0, 0, \cdots, 0, y)$ begins with a term of degree n. Then f is factorable in the form

$$f(x_1, x_2, \cdots, x_m, y) = (y^n + a_1 y + \cdots + a_n)\varphi(x_1, x_2, \cdots, x_m, y),$$

where $a_1, a_2, \cdots, a_n$ are convergent power series in $x_1, x_2, \cdots, x_m$ which vanish for $x_1 = x_2 = \cdots = x_m = 0$, and φ is a power series in $x_1, x_2, \cdots, x_m, y$ which has a constant term different from zero.

In the *Bulletin de la Société Mathématique de France*[*] Goursat has called attention to the fact that the proof which Weierstrass gave of this important theorem, as well as the later proofs which occur in the literature[†], make use of the notions of the function theory, while the theorem itself is essentially of an algebraic character. In the paper referred to he has given an elegant and elementary proof of the theorem which is in outline as follows:

By means of the substitution

$$y^n = - a_1 y^{n-1} - a_2 y^{n-2} - \cdots - a_n$$

the series f can be reduced to a polynomial P of degree $n - 1$ in y, whose n coefficients are convergent series in $a_1, a_2, \cdots, a_n, x_1, x_2, \cdots, x_m$. By the usual theorems of implicit function theory it is shown that the n equations found by putting these coefficients equal to zero have unique solutions for $a_1, a_2, \cdots, a_n$ as power series in $x_1, x_2, \cdots, x_m$, which vanish with $x_1, x_2, \cdots, x_m$. If the values so found are substituted in the formula

$$y^n = - a_1 y^{n-1} - a_2 y^{n-2} - \cdots - a_n + \mu$$

---

[*] "Démonstration élémentaire d'un théoreme de Weierstrass," vol. 36 (1908), p. 209.

[†] See, for example, Picard, Traité d'Analyse, vol. 2, p. 243; Goursat, Cours d'Analyse, vol. 2, p. 284.

5

and the series f again reduced, a polynomial P_1 of degree $n - 1$ in y will be found whose coefficients are series in $x_1, x_2, \cdots, x_m, \mu$. On account of the way in which the functions $a_1, a_2, \cdots, a_n$ were determined, this polynomial P_1 has a factor μ, and hence f has a factor $(y^n + a_1 y^{n-1} + \cdots + a_n)$.

Since the paper of Goursat appeared two further proofs of the theorem have been published, one by the writer* and the other by MacMillan,† each of which seems even more direct than that of Goursat. In the proof which follows use is made of the very concise and elegant method of MacMillan for determining the coefficients, while the rest of the proof is similar to that of the earlier paper of the writer cited above.

The theorem may be stated in a different form as follows:

Suppose that $f(x_1, x_2, \cdots, x_m, y)$ is a series with literal coefficients such that $f(0, 0, \cdots, 0, y)$ begins with the term $a_0 y^n$. Then there is one and but one series $b(x_1, x_2, \cdots, x_m, y)$ which satisfies formally the relation

$$(7) \qquad\qquad bf = p,$$

where p is a polynomial

$$p = a_0 y^n + a_1 y^{n-1} + \cdots + a_n$$

whose coefficients $a_k(x_1, x_2, \cdots, x_m)(k = 1, 2, \cdots, n)$ are series vanishing with the x's.

Each of the coefficients in b and the a's is a rational function of a finite number of the coefficients of f with denominator a power of a_0, and the constant term in b is unity.

If the coefficients in f are chosen numerically so that f converges and $a_0 \neq 0$, then the series b and a_k $(k = 1, 2, \cdots, n)$ also converge.

The functions f, b, p may be written in the forms

$$
\begin{aligned}
f &= a_0 y^n - y^{n+1} f_0 - f_1 - f_2 - \cdots, \\
(8) \qquad b &= b_0 + b_1 + b_2 + \cdots, \\
p &= a_0 y^n - p_1 - p_2 - \cdots,
\end{aligned}
$$

---

* *Bulletin of the American Mathematical Society*, vol. 16 (1910), p. 356.
† Ibid., vol. 17 (1910), p. 116.

where f_k, b_k, p_k are homogeneous expressions of degree k in $x_1, x_2, \cdots, x_m$ with coefficients which are series in y. It is desired to determine b so that the identity (7) holds, and so that the expressions p_k have coefficients which contain y only to the degree $n - 1$.

By substituting the expressions (8) in the identity (7) and equating terms of the same degree in the x's, it follows that

$$b_0(a_0 - yf_0)y^n = a_0y^n,$$
$$b_1(a_0 - yf_0)y^n = b_0f_1 - p_1,$$
$$b_2(a_0 - yf_0)y^n = b_0f_2 + b_1f_1 - p_2,$$
$$\cdot \quad \cdot \quad \cdot \quad \cdot \quad \cdot \quad \cdot \quad \cdot,$$
$$b_k(a_0 - yf_0)y^n = b_0f_k + b_1f_{k-1} + \cdots + b_{k-2}f_2 + b_{k-1}f_1 - p_k,$$
$$\cdot \quad \cdot \quad \cdot \quad \cdot \quad \cdot \quad \cdot \quad \cdot \quad \cdot \quad \cdot \quad \cdot \quad \cdot$$

These equations are to be identities in x and y. The first one determines b_0 uniquely with constant term unity, and furthermore so that each coefficient is a quotient, in fact a polynomial with positive integral coefficients in a finite number of the coefficients of f, divided by a power of a_0. In the second equation p_1 must be chosen equal to the terms of b_0f_1 which contain y to the degree $n - 1$ or less, after which b_1 is uniquely determined. Similarly in the kth equation p_k must first be chosen to cancel the terms on the right of degree $n - 1$ or less in y, and then b_k is unique.

It only remains to show that the series b and a_k are convergent in any numerical case for which f converges. There is no loss of generality in assuming that the series f converges in the domain

$$|x_i| \leqq 1, \qquad |y| \leqq 1 \qquad (i = 1, 2, \cdots, m),$$

since this can always be effected by a substitution of the form

$$x_i = \rho_i x_i', \qquad y = \sigma y' \qquad (i = 1, 2, \cdots, m).$$

Suppose then that K is a number greater than the absolute value of any term in the series $f(1, 1, \cdots, 1, 1)$, that is, greater than the absolute value of any coefficient in f. If A_0 is the absolute value of a_0, the series

$$F = A_0 y^n - \frac{K y^{n+1}}{1 - y} - \frac{K}{1 - y} X,$$

where

$$X = \frac{1}{(1 - x_1)(1 - x_2) \cdots (1 - x_m)} - 1,$$

dominates f in the sense that every coefficient except the first has a numerical value equal to or greater than K; and the series B satisfying the relation

$$BF = A_0 y^n + A_1 y^{n-1} + \cdots + A_n$$

analogous to (7) has coefficients numerically greater than the absolute values of those of b. Hence if B converges the same will be true of b.

But it is easy to show that the series B converges. It will certainly do so if convergent series A_k, C, D can be found satisfying the relation

$$A_0 y^n (1-y) - K y^{n+1} - KX = (A_0 y^n + A_1 y^{n-1} + \cdots + A_n)(Cy + D),$$

because then B would have the value

$$B = \frac{1 - y}{Cy + D}.$$

On comparing the coefficients of the two highest terms in y in the next to last equation, and for convenience denoting by α the constant value

$$\alpha = - \frac{A_0 + K}{A_0^2},$$

it is found that

$$C = \alpha A_0, \qquad D = 1 - \alpha A_1.$$

By comparing the other powers of y and substituting these values, we have

$$A_1 + \alpha A_0 A_2 \qquad\qquad = \alpha A_1^2,$$
$$A_2 + \alpha A_0 A_3 \qquad\qquad = \alpha A_1 A_2$$

$$\cdots\cdots$$

$$A_{n-1} + \alpha A_0 A_n = \alpha A_1 A_{n-1},$$
$$A_n = \alpha A_1 A_n - KX.$$

But these equations have linear terms in A_1, A_2, $\cdots$, A_n with functional determinant different from zero, and hence have solutions, by the theorems of § 2, which are convergent series in x_1, x_2, $\cdots$, x_m and have no constant terms.

It is evident, in any numerical case for which f is convergent, that a neighborhood of the origin may be chosen in which the series b is everywhere different from zero. In such a neighborhood all of the values $(x_1, x_2, \cdots, x_m, y)$ which make f vanish are roots of the equation $p = 0$, and vice versa.

If $f(x_1, 0, \cdots, 0, y)$ has its terms of lowest degree homogeneous and of degree n, then the polynomial $p(x_1, 0, \cdots, 0, y)$ has the same initial terms, since the first coefficient of the factor series b is unity.

§ 10. The Zeros of $\varphi(u, v)$, $\psi(u, v)$, or their Functional Determinant

Consider a function $\varphi(u, v)$ whose values in the neighborhood of the origin in the uv-plane are given by a convergent series in u and v which vanishes for $u = v = 0$. If the series contains a factor u in every term it may be written in the form

$$(9) \qquad \varphi(u, v) = au^k \Phi(u, v),$$

where a is a constant different from zero and $\Phi(u, v)$ is a convergent series for which $\Phi(0, v)$ has a first term of the form v^m with coefficient unity. According to the results of the preceding section, all of the roots of $\Phi(u, v)$ in a neighborhood of the origin will be roots of a certain polynomial

$$(10) \qquad P = v^m + a_1 v^{m-1} + \cdots + a_m,$$

where the coefficients a_k are series in u having no constant terms.

The polynomial P may be equal to the product of two polynomials of similar form,

$$b_0 v^k + b_1 v^{k-1} + \cdots + b_k,$$
$$c_0 v^{m-k} + c_1 v^{m-k-1} + \cdots + c_{m-k},$$

where the coefficients b and c are convergent series in u. In that case the product $b_0 c_0$ must be identically unity, and by dividing the first polynomial by b_0 and multiplying the second by the same series, the two factors will have the form

$$v^k + b_1' v^{k-1} + \cdots + b_k',$$
$$v^{m-k} + c_1' v^{m-k-1} + \cdots + c_{m-k}',$$

The coefficients b' and c' are now series in u without constant terms. Otherwise the product P would have a term of lower degree than v^m, with a coefficient series whose constant term would be different from zero.

It is readily seen from this that the polynomial P is either irreducible in the sense that it can not be decomposed into a product of polynomials of the same sort, or else it is the product of a number of irreducible polynomials of lower degree.

Suppose that $Q(u, v)$ is a polynomial of the form (10) which is irreducible in the sense just described. Then its discriminant with respect to v is a series in u which does not vanish identically, since otherwise Q and Q_v would necessarily have a common factor of the form (10), and Q would not be irreducible. There is a neighborhood $0 < u \leq u_1$ in which the discriminant is everywhere different from zero, and for any value u satisfying these inequalities the values of v making $Q = 0$ are all distinct. According to the results which have been stated above in the introduction to this chapter of the lectures, the values of v which make Q vanish for different values of u will be defined by m series of the form

$$(11) \qquad v = \alpha u^{\mu/p} + \alpha' u^{\mu'/p} + \cdots;$$

and these series must all be distinct, since for sufficiently small values $u \neq 0$, as has been seen, the roots of Q are all distinct.*

It is evident then that all the roots of $\varphi(u, v)$ in the neighborhood of the origin, including those which correspond to the factor u^k in equation (9), are given by a finite number of elements of the form

$$u = at^p, \qquad v = bt^\mu + b't^{\mu'} + \cdots,$$

where a and b do not vanish simultaneously, and p, μ, μ', $\cdots$ are positive integers having no common factor.

The product of factors of the form

(12) $$\{v - \alpha u^{\mu/p} - \alpha' u^{\mu'/p} - \cdots\},$$

corresponding to the elements of a cycle, is a polynomial $Q_1(u, v)$ of the form (10). For the product Q_1 is a series in $u^{1/p}$ and v which is unchanged when $u^{1/p}$ is replaced by $\omega^\nu u^{1/p}$, and Q_1 must therefore contain only powers of $u^{1/p}$ whose exponents are multiples of p, that is, positive integral powers of u.

On the other hand an irreducible polynomial Q possesses only a single cycle of elements of the form (12). Each element of a cycle belonging to Q gives rise, in fact, to a factor Q_1 of Q of the form (10). The number of elements in the cycle could not be greater than the degree of Q, and neither could it be less, since according to the argument of the paragraph just preceding, Q would then be divisible by a factor of the same form corresponding to the product of the factors (12) belonging to the cycle.

By combining these two results, it follows that *the product of the factors of the form* (12) *corresponding to the elements of a single cycle is an irreducible polynomial of the form* (10), *and conversely the elements of an irreducible polynomial of the form* (10) *form a single cycle.*

The Weierstrassian polynomial P of any function φ is a product of irreducible factors of the same form, some perhaps repeated,

---

* The method of proof for this statement in the case of a polynomial P is precisely that of the theory of algebraic functions. See the reference above (page 44) to Appell and Goursat.

to each of which there corresponds a cycle of elements. By the order of an element of φ is meant the number of times its factor (12) is repeated in the product $u^k P$. The order is evidently equal to the multiplicity in $u^k P$ of the irreducible factor to which the element belongs. If φ possesses one element of a cycle it must possess the whole cycle. For the polynomial P belonging to φ has then a common factor with the irreducible polynomial Q of the cycle, and so must be divisible by Q.

Suppose now that $\varphi(u, v)$ and $\psi(u, v)$ are two functions of the form described above, and that the functional determinant

$$(13) \qquad D(u, v) = \begin{vmatrix} \varphi_v & \varphi_v \\ \psi_u & \psi_v \end{vmatrix}$$

does not vanish identically.

If φ and ψ have an element in common, then they have in common the irreducible polynomial Q of the form (10) to which the element belongs, and Q is also factor of D.

The first part of this statement follows from the preceding paragraphs, so that φ and ψ may be supposed to have the forms

$$\varphi = QA, \qquad \psi = QB.$$

When these expressions are substituted in the functional determinant (13) the presence of the factor Q is at once evident.

A similar argument shows that *if φ has an element with corresponding factor Q of multiplicity k, and ψ has the same element and factor with multiplicity l, then D contains the element and its factor with multiplicity $k + l - 1$ at least.*

There is a sort of converse to these statements to the effect that *when φ and D have an element and its factor Q in common, then the element and Q are either multiple in φ or else are common to φ and ψ.*

To prove this let

$$\varphi = QA, \qquad D = QC,$$

and suppose Q not a multiple factor of φ. Then

$$\begin{vmatrix} Q_u A + Q A_u & Q_v A + Q A_v \\ \psi_u & \psi_v \end{vmatrix} = QC;$$

and it follows readily that the determinant

(14)
$$\begin{vmatrix} Q_u & Q_v \\ \psi_u & \psi_v \end{vmatrix}$$

has the factor Q, since A can not have any element in common with Q. Otherwise it would contain the whole irreducible factor Q.

Since Q is irreducible, its discriminant, a series in u, can not vanish identically, and there is an interval $0 < u \leq u_1$ in which it is different from zero. For any value of u satisfying these inequalities the polynomials Q and Q_v have no common root. If

(15) $$u = at^p, \qquad x = \alpha t^\mu + \alpha' t^{\mu'} + \cdots.$$

is the parametric form of one of the elements of Q, then $Q(u, v)$ vanishes identically in t when these expressions are substituted, and $Q_v(u, v)$ is not identically zero in t along the element. Hence there is an interval $0 < t \leq t_1$ in which Q_v is different from zero. Since the determinant (14) has the factor Q and therefore vanishes identically along the curve (15), it follows that

$$Q_v \left(\psi_u \frac{du}{dt} + \psi_v \frac{dv}{dt} \right) \equiv \psi_v \left(Q_u \frac{du}{dt} + Q_v \frac{dv}{dt} \right) \equiv 0$$

is an identity in t. Evidently $\psi(u, v)$ must be constant along the element, and its value is everywhere zero since it vanishes for $t = 0$. Hence ψ has the element (15) in common with Q, and must have Q itself as a factor since Q is irreducible.

The real points (u, v) where one or another of the functions φ, ψ, D vanishes play an important rôle in the investigation which follows. In the discussion of them which follows it will always be understood that when u is real and positive the symbol $u^{1/p}$ stands for the real and positive pth root of u.

If the function φ has no factor u, and if each of its elements when written in the form

(16) $$v = u^{\mu/p}\{\alpha + \alpha'u^{(\mu'-\mu)/p} + \cdots\}$$

has at least one imaginary coefficient, then in a neighborhood of the origin no real point (u, v) with $u > 0$ satisfies the equation $\varphi(u, v) = 0$.

To show this, suppose for the moment that α is imaginary. Then for sufficiently small positive values of u the absolute value of $\alpha'u^{(\mu'-\mu)/p} + \cdots$ will be less than the absolute value of the imaginary part of α, and the parenthesis in the expression (16) will also be imaginary. A similar argument would show v to be complex if one of the higher coefficients were the first not real.

On the other hand, if the coefficients in the expression are all real, then for positive values of u the values of v are real, and the points (u, v) so defined lie on a real arc of the form

$$u = t^p, \qquad v = \alpha t^\mu + \alpha' t^{\mu'} + \cdots \qquad (0 \leqq t \leqq t_1).$$

If the elements of φ are written in the form

(17) $$v = \alpha \epsilon^\mu(-u)^{\mu/p} + \alpha'\epsilon^{\mu'}(-u)'^{\mu/p} + \cdots,$$

where ϵ is a fixed pth root of -1, then an argument similar to that just given shows that $\varphi = 0$ is satisfied by no real points in the neighborhood of the origin with negative values of u, unless at least one of the expressions (17) in $(-u)^{1/p}$ has all of its coefficients real. On the other hand any such element with real coefficients defines points (u, v) on a real arc

$$u = -t^p, \qquad v = \beta t^\mu + \beta' t'^\mu + \cdots \qquad (0 \leqq t \leqq t_1).$$

By combining these results it follows that *all of the real points, in a neighborhood of the origin, which satisfy $\varphi(u, v) = 0$, are the points of a finite number of distinct elements of the form*

(18) $$u = at^p, \qquad v = bt^\mu + b't'^\mu + \cdots \qquad (0 \leqq t \leqq t_1)$$

whose coefficients are real and such that a and b are not both zero.

It may be of interest to note in passing that if an element of φ of the form (16) has real coefficients, then the irreducible polynomial Q which belongs to that element is real. For Q is the product of

$$\{v - \alpha u^{\mu/p} - \alpha' u^{\mu'/p} - \cdots\}$$

and the other factors which arise from it by replacing $u^{1/p}$ by $\omega^\nu u^{1/p}(\nu = 0, 1, 2, \cdots, p-1)$. The coefficients of the product are therefore rational integral functions with real coefficients in the α's and the pth roots of unity, and symmetric in the latter. But symmetric functions of the pth roots of unity are real. A similar remark holds true for the real elements of the form (17).

Two real elements of the form (18) are said to be distinct if there is an interval $0 < t \leqq t_1$ on which the points (u, v) which they define are all distinct. Any two elements are either distinct or else coincident throughout.

Let the two elements have the equations

$$u = at^p, \qquad v = bt^\mu + b't^{\mu'} + \cdots \qquad (0 \leqq t \leqq t_1),$$
$$u = ct^q, \qquad v = dt^\nu + d't^{\nu'} + \cdots \qquad (0 \leqq t \leqq t_2).$$

If $a = c = 0$ then the elements are distinct unless b and d have the same sign, in which case each defines the same half ray from the origin along the v-axis. If $a = 0$, $c \neq 0$ the elements are distinct. If a and c are both different from zero then the elements are distinct unless the expressions

$$v = b\left(\frac{u}{a}\right)^{\mu/p} + b'\left(\frac{u}{a}\right)^{\mu'/p} + \cdots,$$
$$v = d\left(\frac{u}{c}\right)^{\nu/p} + d'\left(\frac{u}{c}\right)^{\nu'/p} + \cdots,$$

are identical in fractional powers of u, in which case the two elements coincide.

It can readily be seen that if two functions φ and ψ have a real element in common then they must each contain the irreducible real factor which belongs to the element.

§ 11. SINGULAR POINTS OF A REAL TRANSFORMATION OF TWO
VARIABLES

In this section it is proposed to study the singular points of a
transformation

(19) $$x = \varphi(u, v), \qquad y = \psi(u, v)$$

for which φ and ψ are convergent series in u, v with real coefficients. It is presupposed that the functional determinant D
of φ and ψ does not vanish identically, and that the real elements
of φ and ψ described in § 10 are all distinct. There is an interval
$0 \leqq t \leqq t_1$ for which the elements of φ, ψ, and D which are
distinct have only the point $(u, v) = (0, 0)$ in common. Some
of these elements may belong to both φ and D, or to ψ and D,
but none are common to φ and ψ. By further restricting the
interval if necessary, it can be effected that the radius

$$\rho = \sqrt{u^2 + v^2}$$

constantly increases on each element as t increases from 0 to t_1.
For ρ is a series in t which does not vanish identically, and its
derivative has the same character. An interval $0 < t \leqq t_1$ can
therefore always be selected on which both ρ and $d\rho/dt$ remain
greater than zero.

It follows immediately that a constant ρ_1 can be selected so
that any circle about the origin of radius ρ_1 or less is intersected
once and but once by each of the elements in question. The
real elements of φ, ψ, and D may therefore be represented as
shown in Fig. 4.

*If the value of ρ_1 is properly restricted then any one of the regions
S shown in the figure is transformed in a one-to-one way by the
equations (19) into a region Σ adjoining the origin and lying
entirely in one quadrant of the xy-plane. The single-valued inverse
functions*

(20) $$u = f(x, y), \qquad v = g(x, y)$$

so defined are continuous over all of Σ and analytic in its interior.

To prove this consider the functions $r(u, v)$ and $\omega(u, v)$ defined by the equations

$$ r = \sqrt{\varphi^2 + \psi^2}, \quad \cos \omega = \frac{\varphi}{r}, \quad \sin \omega = \frac{\psi}{r}. $$

if the radius ρ_1 is properly restricted, then r and ω (modulus 2π) are well defined at every point of the circle with the exception of the origin, since φ and ψ have no real roots in common aside from $(u, v) = (0, 0)$.

The value of r increases monotonically along any analytic curve

$$ u = a_1 t + a_2 t^2 + \cdots, \qquad v = b_1 t + b_2 t^2 + \cdots, $$

for which u and v are not identically zero, as may be seen by reasoning similar to that applied above for ρ, after noting that the series for φ and ψ can not vanish identically in t. In particular

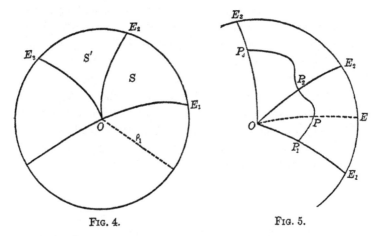

FIG. 4.   FIG. 5.

if ρ_1 is sufficiently small, then r has this property along the boundaries OE_1 and OE_2 of S, and along an auxiliary arc OE chosen arbitrarily for purposes of proof between the two elements OE_1 and OE_2.

Suppose now that k_1 is the minimum of r along the arc E_1E_2, and select arbitrarily a value k between 0 and k_1. The first of the equations

(21) $r(u, v) = k,$ $\omega(u, v) = z$

is satisfied at a unique point $P(u_0, v_0)$ on the arc OE, and the corresponding value of z may be denoted by z_0. The functional determinant of r and ω has the value

$$\frac{\partial(r, \omega)}{\partial(u, v)} = \frac{D(u, v)}{r}$$

and does not vanish anywhere in the interior of S.

The domain in which the equations (21) are to be studied is that consisting of points (u, v, z) for which (u, v) is in S, and z has any real value. According to the first theorem of § 5 and the results of § 2 the equations (21) define two analytic functions

(22) $u = u(z),$ $v = v(z)$

which take the initial values u_0, v_0 when $z = z_0$, and which may be continued over an interval $z_0 \leq z < \zeta''$, as described in § 5. If ζ'' is the value defining the largest such interval, the points $(u(z), v(z))$ corresponding to interior points of the interval will all be interior to S, while as z approaches ζ'' the only limit points of the values $(u(z), v(z))$ must lie on the boundary of S. Otherwise the curve (22) could be continued beyond the value ζ''.

The length of the interval $z_0 \leq z < \zeta''$ is certainly less than $\pi/2$, since in the region S neither $\sin \omega$ nor $\cos \omega$ can vanish. The curve (22) can not intersect itself, since the same values of (u, v) must define the same z by means of the second of equations (21).

As z approaches ζ'', the point $(u(z), v(z))$ approaches a unique limiting point on OE_1 or OE_2. This follows because at any limit point the value of $r(u, v)$ would have to be k, and this can happen at one point P_1 only of PE_2, and at one point P_2 only of OE_2. The curve could not have both P_1 and P_2 as limit points as z approaches ζ'', since then it would necessarily cross the arc OE at the only point P where $r(u, v) = k$, and so would intersect itself.

A similar argument shows that the equations (21) define an arc without double point over an interval $\zeta' < z \leqq z_0$, joining P with that one of the points P_1, P_2 which was not the end of the first arc. For convenience it may be assumed that ζ' is the value belonging to P_1, and ζ'' that for P_2. The preceding inequalities for z would only be reversed if the opposite were the case.

There are no other points in the region S at which $r(u, v) = k$ besides those of the arc P_1P_2 which has just been defined. If there were one not on P_1P_2, it would give rise to a second curve of the same sort joining P_1P_2. But this new curve would necessarily intersect the arc OE at P, and hence must coincide with the original arc P_1P_2 throughout.

For any value $k' < k$ there is a curve similar to P_1P_2 on which all of the points (u, v) making $r(u, v) = k'$ lie.

By means of these results it can now be shown that any two distinct points of the region OP_1P_2 are transformed into two distinct points of the xy-plane. For if (u', v') and (u'', v'') defined the same point (x', y') they would both give $r = \sqrt{x^2+y^2}$ the same value k', and hence must lie on the same curve P_1P_2. But in that case the values of ω corresponding to the two points would necessarily be different, as has been seen above, and hence (x', y') and (x'', y'') could not be the same.

From the final theorem of § 8 it follows at once that the theorem last stated above is true, provided that the circle of radius ρ_1 is altered so that the arc of it which lies between the branches OE_1 and OE_2 lies also within the region OP_1P_2. The region into which S is transformed must lie entirely in one quadrant of the xy-plane, since the values of ω which correspond to points of S are all in one quadrant. In the interior of the image of S the inverse functions (20) are analytic, since at interior points of S the determinant D is different from zero.

Some conclusions with regard to the distribution of the elements of φ, ψ, and D can be readily derived from the discussion just preceding. For example, no region S can be bounded

by two elements of φ. If it were not so, then in a region bounded
by two elements of φ the value of ω on the branch OE_1 would
be everywhere $\pi/2$, or else everywhere $-\pi/2$, and the same is
true for OE_2. But this is impossible since along the arc P_1P_2
the value of ω varies monotonically through an interval less
than $\pi/2$. A similar remark holds for the elements of ψ. Hence
it follows easily that

*Between any elements of D the elements of φ and ψ, if there are
any, must separate each other.*

If the determinant D has opposite signs in two adjoining
regions S and S' of the circle of radius ρ_1 in the uv-plane, shown
in Fig. 5, their transforms in the xy-plane will be folded over
the image of the curve OE_2 and will overlap. In order to prove
this, let it first be remembered that along the element OE_2

$$\frac{dr}{dt} = r_u \frac{du}{dt} + r_v \frac{dv}{dt} \neq 0,$$

so that r_u and r_v can not vanish at any point P_2 different from
the origin. Neither can they vanish at an interior point of
one of the regions S, since at a point where

$$r_u = \frac{\varphi\varphi_u + \psi\psi_u}{r} = 0, \qquad r_v = \frac{\varphi\varphi_v + \psi\psi_v}{r} = 0,$$

the determinant D would necessarily have the value zero, and
this does not occur in the interior of S. The equations

$$r_u \frac{du}{dz} + r_v \frac{dv}{dz} = 0, \qquad \omega_u \frac{du}{dz} + \omega_v \frac{dv}{dz} = 1$$

are satisfied everywhere between P_1 and P_2 on the arc (22).
Hence

$$\frac{du}{dz} = -\frac{r}{D} r_v, \qquad \frac{dv}{dz} = \frac{r}{D} r_u.$$

As z approaches ζ'' the direction cosines of the tangent to the curve (22), for increasing z, approach the values

$$-\frac{r_v}{\sqrt{r_u{}^2+r_v{}^2}}\,, \qquad \frac{r_u}{\sqrt{r_u{}^2+r_v{}^2}}$$

on one of the arcs P_1P_2 and P_2P_3; on the other the limiting direction is exactly the opposite, since the values of D on the two arcs have opposite signs. Hence if $\omega = z$ increases along the arc P_1P_2 it must decrease along P_2P_3, and vice versa.

In the xy-plane these results mean that the images of the arcs P_1P_2 and P_2P_3 are two arcs of the circle $r = k$ which overlap near the image of P_2; the images of S and S' must therefore be superposed in the vicinity of the image of OE_2.

If the boundary OE_2 between S and S' is not one of the elements of D, the images of the two regions in the xy-plane will adjoin each other along the image of OE_2, and the inverse functions (20) will be analytic at every point of the image of OE_2 except the origin. For at such points the functional determinant D is different from zero.

By combining the results which have so far been deduced, the truth of the following theorem is established:

For a transformation

$$(23) \qquad x = \varphi(u, v), \qquad y = \psi(u, v)$$

with the characteristics described in the first paragraph of this section, a circle C can be selected in the uv-plane with center at the origin and having the following properties: The circle is intersected by each real element of the functional determinant D at some first point P. The arcs OP so determined on the different elements divide the interior of C into regions S_1, S_2, $\cdots$, S_k. The points of each region S correspond in a one-to-one way by means of equations (23) with the points of a sheet Σ of the xy-plane which winds about the origin and is bounded by the images of the boundaries of S. The single-valued functions

$$(24) \qquad u = f(x, y), \qquad v = g(x, y)$$

6

so determined are continuous at all points of the sheet Σ and analytic in the interior of Σ. If in two adjoining regions, say S_1 and S_2, the signs of D are opposite, then the images Σ_1 and Σ_2 overlap in the neighborhood of their common boundary $0\pi_2$; if the signs of D are the same, the regions Σ_1 and Σ_2 adjoin along $0\pi_2$ without over-lapping.

The adjoining figure illustrates the case when D has four real elements and the signs of D are opposite in any two adjoining regions S. Further illustrations of the theorem are given in § 14.

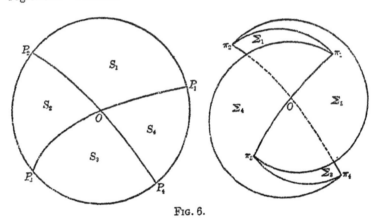

Fig. 6.

It has not been proved above that the functions (24) are continuous on a boundary 0π of one of the regions Σ. Suppose that π is a point of such a boundary, and let

(25) $\pi_1, \ \pi_2, \ \pi_3, \ \cdots$

be any sequence of points of Σ with limit π. The corresponding points

(26) $p_1, \ p_2, \ p_3, \ \cdots$

of S have condensation points in S, one of which may be denoted by p. There is then a sub-sequence

$$p_1', \ p_2', \ p_3', \ \cdots$$

among the points (26) whose limit is p; and on account of the continuity of the functions (23), the corresponding points

$$(27) \qquad \pi_1', \ \pi_2', \ \pi_3', \ \cdots$$

of the sequence (25) must have as limit point the image of p in Σ. But the limit of (27) is necessarily π, and π is therefore the image of p. It follows at once that the sequence (26) has a unique limit point p which is the image of π, and from this property the continuity of the functions (24) in the ordinary sense can be readily deduced.

The functions φ, ψ, and D can be expanded in the form

$$(28) \qquad \begin{aligned} \varphi &= \varphi_m + \varphi_{m+1} + \cdots, \\ \psi &= \psi_n + \psi_{n+1} + \cdots, \\ D &= D_{m+n-2} + D_{m+n-1} + \cdots, \end{aligned}$$

where φ_k, ψ_k, D_k are homogeneous polynomials in u, v of degree k, and

$$D_{m+n-2} = \begin{vmatrix} \dfrac{\partial \varphi_m}{\partial u} & \dfrac{\partial \varphi_m}{\partial v} \\[2mm] \dfrac{\partial \psi_n}{\partial u} & \dfrac{\partial \psi_n}{\partial v} \end{vmatrix}$$

If the real roots of φ_m, ψ_n, and D_{m+n-2} are all simple roots and distinct from each other, there will be an element of φ, ψ, or D in each of the corresponding directions, and a notion of the character of the transformation can be derived without difficulty. In the applications of § 14 this remark is of frequent service.

§ 12. The Case where the Functional Determinant Vanishes Identically

It is well known that when the functional determinant of two analytic functions φ and ψ vanishes identically, then near any point where not all of the derivatives $\varphi_u, \varphi_v, \psi_u, \psi_v$ vanish the functions φ and ψ satisfy a relation of the form

$$F(\varphi, \psi) = 0$$

identically in v and v. It is possible to show that such a relation exists also near a singular point at which the four derivatives above all vanish.

If a relation can be found after a substitution of the form

$$(29) \qquad u = \alpha u_1 + \beta v_1, \qquad v = \gamma u_1 + \delta v_1,$$

for which $\alpha\delta - \beta\gamma$ does not vanish, then it will surely be satisfied when u_1 and v_1 are replaced by the original variables u, v.

Suppose then that the analytic functions φ and ψ have already been prepared by a transformation (29) in such a way that in the expansions (28) φ_m and ψ_n both contain terms in u alone. By applying the preparation theorem of Weierstrass to the functions $\varphi(u, v) - x$ and $\psi(u, v) - y$, two polynomials

$$P(u, v, x) = u^m + a_1 u^{r-1} + \cdots + a_m,$$
$$Q(u, v, y) = u^n + b_1 u^{n-1} + \cdots + b_n$$

are obtained, whose coefficients are convergent series, without constant terms, in v, x and v, y, respectively. In a certain vicinity

$$|x| < \epsilon, \quad |y| < \epsilon, \quad |u| < \epsilon, \quad |v| < \epsilon$$

the only solutions of the equations

$$(30) \qquad \varphi(u, v) - x = 0, \qquad \psi(u, v) - y = 0$$

are values (u, v, x, y) which make P and Q vanish also, and vice versa.

The resultant of P and Q is a convergent series $R(v, x, y)$ for which $R(0, x, y)$ does not vanish identically. For if all of the coefficients of $R(0, x, y)$ were zero, there would be a region

$$(31) \qquad v = 0, \quad |x| < \delta, \quad |y| < \delta \qquad\qquad (\delta \leqq \epsilon)$$

at any point of which the polynomials P and Q have a common root in absolute value less than ϵ, and the set of values $(u, 0, x, y)$ so defined satisfies also the equations (30). The existence of such a region is, however, impossible, since when y' is given satisfying (31), a value x' can always be selected which is different from the values of $\varphi(u, 0)$ at all of the n roots of $Q(u, 0, y')$. For such a set $v = 0$, x', y' in the region (31) there would be no corresponding value u' satisfying the equations (30).

The resultant $R(v, x, y)$ vanishes identically in u, v when x and y are replaced by φ and ψ. For R is expressible in the form

$$R(v, x, y) = MP + NQ,$$

where M and N are polynomials in u with coefficients which are series in v, x, y, and P and Q vanish identically when $x = \varphi$, $y = \psi$.

The series $R(0, \varphi, \psi)$ vanishes identically in u, v. If not, there would be a straight line $u = kv$ on which $R(0, \varphi, \psi)$ and φ_u are different from zero except at the origin. Let (u', v') be a point of this line near $(u, v) = (0, 0)$, at which φ and ψ have the values φ' and ψ', respectively. The series

(32) $$R(0, \varphi, \psi) + R_v(0, \varphi, \psi)v + \cdots$$

vanishes identically, in particular along the curve

(33) $$\varphi(u, v) = \varphi'$$

through the point (u', v'). Since φ_u does not vanish at (u', v'), this curve can be expressed in the form

$$u = U(v),$$

and along it

$$\frac{d}{dv}\psi(U, v) = \left[\psi_u \frac{-\varphi_v}{\varphi_u} + \psi_v\right]_{u = U(v)} = 0,$$

since the functional determinant of φ and ψ vanishes identically. On the curve (33) the function ψ has therefore the constant value ψ', and the series (32) takes the form

$$R(0, \varphi', \psi') + R_v(0, \varphi', \psi')v + \cdots$$

and vanishes identically in v. Its coefficients must therefore all vanish, since a series whose zeros have a point of condensation in the interior of its circle of convergence must have all of its coefficients equal to zero. This contradicts, however, the assumption that a point (u', v') exists at which $R(0, \varphi, \psi)$ does not vanish.

It has been shown therefore that *in case the functional determinant of the two convergent series*

$$\varphi = \varphi_m + \varphi_{m+1} + \cdots,$$
$$\psi = \psi_n + \psi_{n+1} + \cdots$$

vanishes identically, the two functions φ, ψ satisfy a relation of the form

$$F(\varphi, \psi) = 0$$

identically in u, v, where F is itself a convergent series in its two arguments. This statement is true even when φ and ψ both have singular points at the origin.

It is evident that when $D = 0$ the transformation

$$x = \varphi(u, v), \qquad y = \psi(u, v)$$

makes all of the points in the neighborhood of the origin in the uv-plane correspond to points on the various branches of the curve

$$F(x, y) = 0$$

in the xy-plane. The points (x, y) which are obtained by the transformation do not cover any region.

§ 13. A GENERALIZATION OF THE PREPARATION THEOREM OF WEIERSTRASS

Consider for a moment two functions

$$(34) \qquad f(u, v, x_1, x_2, \cdots, x_m), \quad g(u, v, x_1, x_2, \cdots, x_m)$$

which are polynomials in the variables u, v and have for coefficients convergent series in $x_1, x_2, \cdots, x_m$. According to the usual algebraic theory of elimination, there exists a polynomial p in v

which has convergent series in the x's as coefficients, and which is linearly expressible in the form

$$p = cf + dg,$$

where c and d are polynomials of the same character as f and g. If a set of variables (u, v, x) make f and g both vanish, then v must be a root of the polynomial p; and conversely to any root of p corresponding to given values x, there exists at least one pair of values (u, v) which satisfy the two equations $f = g = 0$.

There is a generalization of the preparation theorem of Weierstrass from which similar results may be deduced with respect to two functions f and g which are not polynomials but series in the variables u and v, and with respect to the roots of such functions in a neighborhood of any set of values (u_0, v_0, x_0) making f and g vanish. As in the proof of the theorem of § 9, the point in whose neighborhood f and g are to be studied may be taken without loss of generality at the origin.

Suppose then that f and g are two convergent series in u, v, x vanishing for $(u, v, x) = (0, 0, 0)$, and such that $f(u, v, 0, 0, \cdots, 0)$ and $g(u, v, 0, 0, \cdots, 0)$ have no common factor. Then there exists a polynomial

$$(35) \qquad p = v^n + p_1 v^{n-1} + \cdots + p_n,$$

in which the coefficients p_k $(k = 1, 2, \cdots, n)$ are convergent series in x having no constant terms, with the following properties: (1) it is linearly expressible in the form

$$p = cf + dg,$$

where c and d are convergent power series in u, v, x; (2) in a properly chosen neighborhood

$$(36) \qquad |u| < \epsilon, \quad |v| < \epsilon, \quad |x| < \epsilon$$

every root (u, v, x) of f and g must also make p vanish; (3) there exists a constant $\delta \leq \epsilon$ such that for any x in the region

$$(37) \qquad |x| < \delta$$

there is associated with each root v of p a solution (u, v, x) of the equations $f = g = 0$ satisfying the inequalities (36).*

If $f(u, v, 0, 0, \cdots, 0)$ and $g(u, v, 0, 0, \cdots, 0)$ have no common factor, then one at least of them, say f, has terms in the variable u alone, and according to the preparation theorem of Weierstrass $f(u, v, x)$ has as factor a polynomial of the form

$$(38) \qquad a_0 u^m + a_1 u^{m-1} + \cdots + a_{m-1} u + a_m = bf,$$

in which a_0 is a constant different from zero, and $a_1, a_2, \cdots, a_m$ are series in v, x without constant terms. The symmetric functions of the roots $u_1, u_2, \cdots, u_m$ of this polynomial are expressible rationally and integrally in terms of the coefficients $a_1, a_2, \cdots, a_m$, and are therefore convergent series in v, x. The product

$$(39) \qquad \prod_{k=1}^{m} g(u_k, v, x) = h(v, x)$$

is a convergent series in u_k, v, x, also symmetric in the variables u_k. and hence expressible as convergent series in v, x.

The function $h(v, 0)$ does not vanish identically, on account of the hypothesis that $f(u, v, 0, 0, \cdots, 0)$ and $g(u, v, 0, 0, \cdots, 0)$ have no common factor. If it did vanish identically, then for every sufficiently small value of v one at least of the expressions $g(u_k, v, 0)$ would vanish. But in § 10 it was seen that when $f(u, v, 0)$ and $g(u, v, 0)$ have no factor in common, there is always an interval $0 < v \leqq v_1$ in which there is no value v belonging to a pair (u, v) making both of these functions vanish.

The preparation theorem of Weierstrass can therefore be applied also to the function $h(v, x)$, and the polynomial so found is the one desired in the theorem. For, in the first place, a constant ϵ can be chosen so small that every root (u, v, x) of f and g in the region (36) must be one of the sets (u_k, v, x), and must make

---

* A proof that the values of u and v belonging to the roots of a system of equations of the form (34) are roots of polynomials similar to (35) was given by Poincaré in the introduction to his Thesis, "Sur les propriétés des fonctions définies par les équations aux différences partielles," Paris (1879).

the product (39), and hence p, vanish. In the second place, a constant $\delta \leq \epsilon$ can be taken so small that every root v of p as well as the corresponding sets (u_k, v, x) lie in the domain (36). One at least of these sets must evidently satisfy $g = 0$ as well as $f = 0$. The restrictions on δ and ϵ have been stated somewhat roughly, but the reader will readily convince himself that these quantities may be selected so that the convergence of the different series and their equivalence with the corresponding polynomials are properly adjusted.

Finally, the polynomial p is linearly expressible in the form described in the theorem, in terms of f and g. To prove this, suppose that the above process has been applied to the functions $f - \alpha$ and $g - \beta$. A polynomial $P(v, x, \alpha, \beta)$ with coefficients which are series in x, α, β is then found, which may be written in the form

$$P(v, x, \alpha, \beta) = P(v, x, 0, 0) + C\alpha + D\beta,$$

where C and D are convergent series in the arguments of P. The series $P(u, x, f, g)$ vanishes identically in u, v, x since $P = 0$ must be satisfied by every set of variables (u, v, x, α, β) in a neighborhood of the origin which make $f - \alpha$ and $g - \beta$ vanish, certainly then by the set (u, v, x, f, g). Hence

$$P(v, x, 0, 0) = - Cf - Dg$$

is an identity in u, v, x, when α and β are replaced in C and D by the series f, g. But $P(v, x, 0, 0)$ is precisely the polynomial $p(v, x)$ found above, since for $\alpha = \beta = 0$ the steps in the construction of $P(v, x, 0, 0)$ are identical with those used in finding p.

If the series $f(u, v, 0, 0, \cdots, 0)$ and $g(u, v, 0, 0, \cdots, 0)$ begin with homogeneous polynomials having no common factor of degrees m and n, respectively, then the degree of the polynomial p is $\nu = mn$.

---

* In a paper of recent date the writer has developed a generalization of this theorem and the results which follow, for a system of equations of the form $f_i(x_1, x_2, \cdots, x_m; y_1, y_2, \cdots, y_n) = 0$ $(i = 1, 2, \cdots, n)$. See *Transactions of the American Mathematical Society*, vol. 13 (1912), p. 133.

Let the lowest terms of $f(u, v, 0, 0, \cdots, 0)$ and $g(u, v, 0, 0, \cdots, 0)$ be denoted by $\varphi_m(u, v)$ and $\psi_n(u, v)$, respectively. One of the two, say φ_m, has a term involving u alone with coefficient different from zero, since φ_m and ψ_n have no common factor. The terms of lowest degree in the polynomial (35) are also φ_m, since the series b has constant term unity. In the product (39) the terms may be rearranged into groups of the form $cv^\rho U$, where U is a homogeneous symmetric function of a certain degree σ in $u_1, u_2, \cdots, u_m$. The expression for such a symmetric function is isobaric and has the weight σ in the coefficients of the polynomial (35). When $x = 0$ the terms of lowest degree in U will be at least of degree σ in v, since each coefficient a_k of (35) begins with the coefficient of u^{m-k} in the polynomial $\varphi_m(u, v)$. The terms of lowest degree in v alone in the product (39) will therefore be those of the product

$$\prod_{k=1}^{m} \psi_n(u_k, v),$$

and they have the value $v^{mn}R/a_0$, in which a_0 is the coefficient of v^m in $\varphi_m(u, v)$ and R is the resultant of $\varphi_m(1, v)$ and $\psi_n(1, v)$.[*] But since φ_m and ψ_n have no common factor the coefficient of v^{mn} is surely different from zero, and the theorem last stated follows at once.

If the substitution

$$v = -\, tu + z$$

is made, in which t is a new variable, the series

$$(40) \qquad \begin{aligned} F(u, z, x, t) &= f(u, z - tu, x), \\ G(u, z, x, t) &= g(u, z - tu, x) \end{aligned}$$

have a polynomial

$$(41) \qquad P(z; x, t) = CF + DG$$

with properties similar to those of p and of the same degree ν. In

---

[*] See, for example, König, Einleitung in die allgemeine Theorie der algebraischen Grössen, p. 311 and p. 271 (d).

a properly chosen region

(42) $$|u| < \epsilon, \quad |v| < \epsilon, \quad |x| < \epsilon$$

every root (u, v, x) of f and g defines a factor $z - tu - v$ of P. If $\delta \leq \epsilon$ is sufficiently small and x a set of variables satisfying

(43) $$|x| < \delta,$$

then P has ν factors of the form $z - tu - v$, for each of which the values (u, v, x) are a solution of the equations $f = g = 0$ in the region (42).

The degree of P must be the same as that of p, since for $x = t = 0$ the series $F(u, \bar{z}, 0, 0)$, $G(u, z, 0, 0)$ are identically equal to the series $f(u, v, 0)$ and $g(u, v, 0)$ when v is replaced by z. In a certain region

(44) $$|u| < \epsilon_1, \quad |z| < \epsilon_1, \quad |x| < \epsilon_1, \quad |t| < \epsilon_1,$$

where ϵ_1 is for convenience taken less than unity, every root system (u, z, x, t) of F and G makes P vanish also. If ϵ is taken less than $\epsilon_1/2$ and t is restricted to the range $|t| < \epsilon_1$, every root system (u, v, x) of f and g in the region (42) gives values u, $z = tu + v$, x, t satisfying the inequalities (44), and hence P must vanish identically in t and have $z - tu - v$ as a factor.

Suppose then that ϵ is a constant satisfying the requirements of the theorem with respect to the region (42), and that the region analogous to (37) for the polynomial P and the constant $\epsilon/2$ is

(45) $$|x| < \delta, \quad |t| < \delta;$$

and let $x = \xi$ be any set of values satisfying these inequalities. If the discriminant of P is not identically zero in t for $x = \xi$, a value $t = \tau$ can be selected also satisfying (45) and such that all the roots z of P corresponding to the values ξ, τ are distinct. There are then ν distinct root systems (u, z, ξ, τ) satisfying the inequalities (41) with ϵ_1 replaced by $\epsilon/2$. The corresponding values $(u, v = z - tu, \xi)$ are ν distinct roots of f and g lying in the region (41). According to the paragraph just preceding, P has therefore ν distinct factors $z - tu - v$.

In case the discriminant of P vanishes identically in t for $x = \xi$, the multiple factors of $P(z; \xi, t)$ can be separated out by the highest common divisor process, and the factorization of the resulting polynomial can then be discussed in a manner similar to that just explained. In either case, therefore, $P(z; \xi, t)$ has only linear factors of the form $z - tu - v$.

The number and character of the root systems (u, v, x) of the functions f and g in the neighborhood of the origin are well defined by means of the polynomial $P(z; x, t)$. To any x in the region (43) there correspond ν root systems (u, v, x) not necessarily all distinct, and the ν-valued functions $u(x), v(x)$ so defined are continuous. This is evidently true for the function $v(x)$, since its values are the roots of the polynomial $P(v; x, 0)$ whose coefficients are analytic in x. Similarly z is continuous in x, t, since its values are the roots of $P(z; x, t)$, and it follows that $u = (z - v)/t$, for a fixed value $t \neq 0$, must be continuous in x.

If P is not irreducible, that is, not decomposable into similar factors of lower degrees, its discriminant $\Delta(x, t)$ can not vanish identically in x, t. At any value $x = \xi$ where $\Delta(\xi, t)$ is not identically zero in t, the ν factors $z - tu - v$ of P are all distinct. If $t = \tau$ is selected so that $\Delta(\xi, \tau) \neq 0$, the roots of P are distinct analytic functions of x and t in the neighborhood of ξ, τ, and the corresponding values of u and v are analytic functions of x in the neighborhood of ξ.

The values $x = \xi$ near which the ν-valued functions u, v do not surely have ν distinct analytic branches, are those for which $\Delta(\xi, t)$ vanishes identically in t. At such a point some of the values of the root-systems (u, v) coincide, and only those which are distinct belong necessarily to analytic branches of the functions u, v. The values ξ which make $\Delta(\xi, t)$ identically zero must belong to one of the totalities of points defined by equating to zero the coefficients of the finite number of powers of t in the discriminant $\Delta(x, t)$.*

---

* For the characterization of these totalities after the method of Kronecker for algebraic equations, see Kistler, "Ueber Funktionen von mehreren komplexen Veränderlichen," Dissertation, Göttingen, 1905.

If $P(z, x, t)$ is reducible, arguments similar to those above can be applied to any one of its irreducible factors.

The multiple roots (u, v, x) of the functions f and g are character-ized by the property that the functional determinant $\partial(f, g)/\partial(u, v)$ is zero at such points.

For from the identity (41) in u, z, x, t, it follows by differentiation that

$$0 \equiv C_u F + D_u G + C F_u + D G_u,$$
(46)
$$P_z \equiv C_z F + D_z G + C F_z + D G_z.$$

If the determinant

$$\begin{vmatrix} F_u & F_z \\ G_u & G_z \end{vmatrix} = \begin{vmatrix} f_u & f_v \\ g_u & g_v \end{vmatrix}$$

vanishes at a solution (u, v, x) of $f = g = 0$, the two equations above show that

$$CF_u + DG_u = 0, \qquad P_z = CF_z + DG_z = 0$$

for the values $(u, z = -tu + v, x)$; and it follows that $z-tu-v$ is a multiple factor of P, since it occurs also in P_z.

On the other hand suppose that at a set of values (u', v', ξ) the determinant $\partial(f, g)/\partial(u, v)$ is different from zero, while f and g vanish. It is to be shown that the polynomial $P(z; \xi, t)$ has $tu' + v'$ as a simple root. All of the roots of $P(z; \xi, t)$ have the form $tu + v$, and some are perhaps multiple. Those which are distinct will remain distinct for a numerical value $t = \tau$ if τ is properly selected, and the derivative

(47)
$$F_y(u', \zeta, \xi, \tau) = f_u(u', v', \xi) - \tau f_v(u', v', \xi)$$

can at the same time be made different from zero, ζ being the expression $\tau u' + v'$. In the expressions

(48)
$$A_0 u^m + A_1 u^{m-1} + \cdots + A_{m-1} u + A_m = BF,$$

(49)
$$\prod_{k=1}^{m} G(u_k, z, \xi, \tau) = H(z, \xi, \tau),$$

analogous to (38) and (39) for the functions $F(u, z, \xi, t)$ and $G(u, z, \xi, t)$, the factor $G(u_1, z, \xi, \tau)$, where u_1 is the root of (48) which reduces to u' for $z = \zeta$, is the only one which vanishes for $z = \zeta$. To prove this it can be seen in the first place that u' is a simple root of (48) for $z = \zeta$, since the derivative (47) is different from zero. Furthermore when $z = \zeta$ no other root u_2 distinct from u_1 can make $G(u_2, z, \xi, \tau)$ vanish. Otherwise f and g would vanish not only at the values (u', v', ξ), but also at $(u_2', \zeta - \tau u_2', \xi)$, where u_2' is the value of u_2 for $z = \zeta$; and $P(z; \xi, t)$ would have two roots, $tu' + v' = tu' + \zeta - \tau u'$ and $tu_2' + \zeta - \tau u_2'$, which are distinct for $t \neq \tau$ and equal to ζ when $t = \tau$. On account of the way in which τ was selected, this is impossible.

The root u_1 of (48), that is to say also of F, has an expansion of the form

$$u_1 - u' = -\frac{F_z(u', \zeta, \xi, \tau)}{F_u(u', \zeta, \xi, \tau)} (z - \zeta) + \cdots$$

in powers of $z - \zeta$; and the value of $G(u_1, z, \xi, \tau)$ is a series

$$\frac{F_u G_z - F_z G_u}{F_u} (z - \zeta) + \cdots$$

whose first term is different from zero, since for the values (u', ζ, ξ, τ) we have

$$\begin{vmatrix} F_u & F_z \\ G_u & G_z \end{vmatrix} = \begin{vmatrix} f_u(u', v', \xi) & f_v(u', v', \xi) \\ g_u(u', v', \xi) & g_v(u', v', \xi) \end{vmatrix} \neq 0,$$

as is readily seen from equations (40). Hence the quotient $H(z, \xi, \tau)/(z - \zeta)$ is different from zero, and neither $H(z, \xi, t)$ nor its polynomial $P(z; \xi, t)$ can have more than one factor $z - tu' - v'$.

§ 14. Applications of the Preceding Theory

The real transformation

$$(50) \qquad \begin{aligned} x &= \varphi(u, v) = a_{10}u + a_{01}v + a_{20}u^2 + \cdots, \\ y &= \psi(u, v) = b_{10}u + b_{01}v + b_{20}u^2 + \cdots \end{aligned}$$

has a singular point at the origin when

(51)
$$\begin{vmatrix} a_{10} & a_{01} \\ b_{10} & b_{01} \end{vmatrix} = 0.$$

If one of the elements of the determinant is different from zero, it may be assumed without loss of generality to be a_{10}; then after two transformations

$$u' = a_{10}u + a_{01}v, \qquad v' = v,$$
$$x' = x, \qquad y' = -\frac{b_{10}}{a_{10}}x + y$$

the equations (50) take the form

(52)
$$x = u + a_{20}u^2 + a_{11}uv + a_{02}v^2 + \cdots,$$
$$y = \qquad b_{20}u^2 + b_{11}uv + b_{02}v^2 + \cdots.$$

For convenience the primes have been dropped, and the notation for coefficients of terms of higher degree than the first is the same as that in the original equation. It may further be supposed that the polynomials

$$\varphi_1 = u, \qquad \psi_2 = b_{20}u^2 + b_{11}uv + b_{02}v^2$$

have no common factor, in other words that $b_{02} \neq 0$. The origin is then a singular point for the transformation (50) of a very general type, since aside from the assumption (51) only inequalities on the coefficients of the series have been exacted.

The functional determinant has the expansion

$$D(u, v) = b_{11}u + 2b_{02}v + \cdots,$$

and hence has a single branch

$$v = -\frac{b_{11}}{b_{02}}u + \cdots,$$

along which D vanishes and on opposite sides of which D has different signs. The image Δ of this curve in the xy-plane has an ordinary point at the origin, as shown by its equations

$$x = u + \cdots, \qquad y = \frac{4b_{02}b_{20} - b_{11}{}^2}{b_{02}}u^2 + \cdots.$$

The region S in the figure has in it one real element of φ and at most two of ψ, since the solutions of $\varphi = 0$ lie on a single real

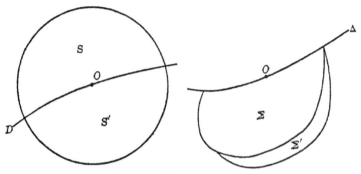

Fig. 7.

curve through the origin, and those of $\psi = 0$ are either imaginary or else lie on two real branches. Hence the region Σ which is the image of S lies on one side only of the curve Δ and overlaps the image Σ' of S'.

Since φ_1 and ψ_2 have no common factor, the theorems of § 13 show that there exist two constants, δ and ϵ, such that the equations (52) have two and only two solutions $[u_1(x, y), v_1(x, y), x, y]$, $[u_2(x, y), v_2(x, y), x, y]$ in the region

$$|u| < \epsilon, \qquad |v| < \epsilon, \qquad |x| < \epsilon, \qquad |y| < \epsilon$$

corresponding to any (x, y) in the region

$$|x| < \delta, \qquad |y| < \delta.$$

The functions u_1, v_1, u_2, v_2 so defined are everywhere continuous and the two solutions above are analytic and distinct except along the curve Δ. On one side of Δ they are imaginary, on the other real.

Another interesting case is that of a transformation (50) for which again the coefficients are real, and

$$\frac{\partial \varphi}{\partial u} = \frac{\partial \psi}{\partial v}, \qquad \frac{\partial \varphi}{\partial v} = -\frac{\partial \psi}{\partial u}.$$

Such a transformation might be called a monogenic transformation. It follows at once that φ and ψ must begin with two homogeneous polynomials, φ_m and ψ_m, of the same degree m, which also satisfy the last equations. Consequently

$$\varphi_m + i\psi_m = (a + ib)(u + iv)^m = \rho^m(a + ib)(\cos \theta + i \sin \theta)^m$$

and

$$\varphi_m = \rho^m(a \cos m\theta - b \sin m\theta), \quad \psi_m = \rho^m(a \sin m\theta + b \cos m\theta),$$

where a and b are not both zero. These equations show that $\varphi_m(u, v)$ and $\psi_m(u, v)$ have each m real linear factors in u, v, and that no factor of φ_m is also in ψ_m.

The determinant $D(u, v)$ has an expansion

$$D(u, v) = D_{2m-1} + D_{2m} + \cdots,$$

where

$$D_{2m-1} = \begin{vmatrix} \dfrac{\partial \varphi_m}{\partial u} & \dfrac{\partial \varphi_m}{\partial v} \\ \dfrac{\partial \psi_m}{\partial u} & \dfrac{\partial \psi_m}{\partial v} \end{vmatrix} = \left(\frac{\partial \varphi_m}{\partial u}\right)^2 + \left(\frac{\partial \varphi_m}{\partial v}\right)^2.$$

The homogeneous polynomial D_{2m-1} has no real root, since such a root would necessarily belong to both $\partial \varphi_m/\partial u$ and $\partial \varphi_m/\partial v$, and from the equations

$$m\varphi_m = u\frac{\partial \varphi_m}{\partial u} + v\frac{\partial \varphi_m}{\partial v}, \quad m\psi_m = u\frac{\partial \psi_m}{\partial u} + v\frac{\partial \psi_m}{\partial v} = -u\frac{\partial \varphi_m}{\partial v} + v\frac{\partial \varphi_m}{\partial u}$$

it follows that φ_m and ψ_m would then have a common factor. Hence there are no real points at which D vanishes near the origin in the uv-plane.

7

The argument of § 11 shows that the elements of φ_m and ψ_m separate each other and that a neighborhood of the origin in the uv-plane is transformed into a sheet winding m times around the origin in the xy-plane, as shown in the figure. This is the

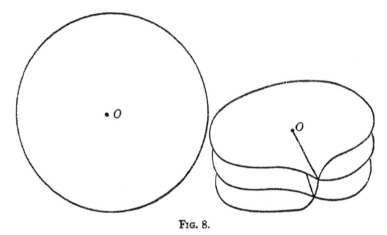

Fig. 8.

well-known transformation of the neighborhood of the origin in a complex w plane by means of a relation of the form

$$= Aw^m + A'w^{m+1} + \cdots,$$

where $z = x + iy$ and $w = u + iv$. The figure is drawn for $m = 3$.

There are many other special cases similar to those just given which might be elucidated by means of the theorems of the preceding sections, but for which the methods in the two examples just given are typical. It may be of interest, however, to exhibit an example which illustrates the use of the theorems of § 8, as well as the behavior of a transformation at singular points.

Suppose that the real uv-plane is transformed by means of the equations

$$(53) \qquad x = \frac{u^2}{2} - uv + \frac{v^2}{2} + \frac{u^3}{3}, \qquad y = \frac{u^2}{2} + uv + \frac{v^2}{2}.$$

The functional determinant has the value

$$D(u, v) = (u + v)(u^2 + 2u - 2v)$$

and it vanishes along the curves

$$v = -u, \qquad v = u + \frac{u^2}{2},$$

which have, respectively, the images

(54)
$$x = 2u^2 + \frac{u^3}{3}, \qquad y = 0,$$

$$x = \frac{u^3}{3} + \frac{u^4}{8}, \qquad y = \frac{u^2}{8}(u + 4)^2$$

in the xy-plane. These curves are shown in the accompanying

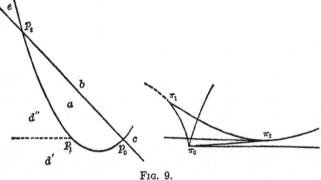

Fig. 9.

figures, the x-axis being drawn triply between $x = 0$ and $x = 32/3$ since this segment is described three times by the point (54) with varying u. To the auxiliary arc $-\infty < u \leqq -2,\ v = 0$ there corresponds the curve

$$x = \frac{u^2}{2} + \frac{u^3}{3}, \qquad y = \frac{u^2}{2} \quad (-\infty < u \leqq -2)$$

shown dotted in the figure.

Consider now, for example, the region a in the uv-plane.

Its boundary is transformed into the boundary of the region α in Fig. 10. According to the generalization of the theorem of

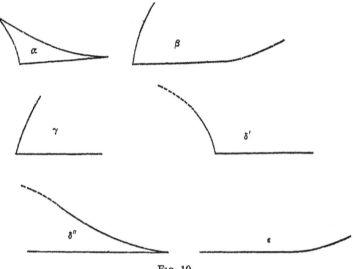

Fig. 10.

Schoenflies in § 8, the transformation defines a one-to-one correspondence between the regions a and α; and the inverse functions $u(x, y)$, $v(x, y)$ so defined are continuous over α and analytic in its interior.

Consider now the region of points (u, v, x, y) defined by the conditions that (u, v) shall lie in the region b or on its boundary, while (x, y) is unrestricted. There is but one sheet of solutions of equations (53) in this region, since any two particular solutions (u', v', x', y'), (u'', v'', x'', y'') interior to the sheet can be joined by a continuous curve lying entirely within the sheet, as may be seen by joining (u', v'), (u'', v'') by a continuous curve in b. No one of the solutions in question has a projection (x, y) outside of β, since otherwise every point exterior to β would be such a projection, according to the third theorem of § 5 or the fourth of § 8; and from the second of equations (53) it is evident that no solution (u, v, x, y) has a negative value for y. On the other hand every point of β is the projection of a solution. Since β is simply connected, it follows from the fourth theorem of § 8 that the sheet of solutions is single-valued and that the equations (53) define a one-to-one correspondence between b and β similar to that for a and α.

A similar argument can be made for each of the regions shown in the figure and its corresponding image in the xy-plane.

CHAPTER III

It is not within the limited scope of these lectures to give a complete account of the various methods for proving the existence of a system of solutions of a set of ordinary differential equations, nor would it be advisable, in view of the many able presentations of these fundamental theorems already well known in mathematical literature. It is rather the intention of the writer to insist on conclusions which can be derived from known methods with regard to the behavior of solutions in any region of size and shape compatible with the continuity properties of the functions by means of which the equations are defined, as over against the usual restriction of the problem to a rectangular or circular neighborhood of a particular point. It has been remarked by Picard* and Painlevé† that if a continuous solution of the differential equation

$$(1) \qquad \frac{dy}{dx} = f(x, y)$$

exists over an interval $\alpha \leqq x \leqq \beta$, then the Cauchy polygons of approximation are defined and converge uniformly to the solution for all values of x in the interval. In § 17 below it is shown that in a region R in which the function f is continuous and satisfies the so-called Lipschitz condition, the polygons of Cauchy passing through a given initial point (ξ, η) interior to R define a priori a continuous solution of the differential equation extending to infinity or else to the boundary of the region. It follows then that there is a function

$$(2) \qquad y = \varphi(x, \xi, \eta)$$

---

* *Comptes Rendus*, vol. 128 (1899), page 1363.

† *Bulletin de la Société Mathématique de France*, vol. 27 (1899), p. 151.

satisfying the differential equation (1) and defined over a region of points (x, ξ, η) of the form

$$(\xi, \eta) \text{ interior to } R, \quad \alpha(\xi, \eta) < x < \beta(\xi, \eta),$$

and as x approaches α or β the only limiting points which the points (x, y) defined by the function (2) can have are at infinity or else on the boundary of the region R.

In § 18 attention is called to the theorems of Bendixon by means of which it can be shown that the function φ is continuous, and in certain circumstances differentiable with respect to the arguments ξ, η as well as with respect to x. The " imbedding theorem " of Bolza* which asserts that any given solution, near which the function f has suitable continuity properties, can be imbedded in a one-parameter family of neighboring solutions of the differential equation, is an immediate consequence of these results, an analogue for differential equations of the fundamental theorem for implicit functions proved in § 1.

The methods mentioned above are applicable almost without change of wording to a system of equations

$$\frac{dy_\beta}{dx} = f_\beta\,(x, y_1, y_2, \cdots, y_n) \quad (\beta = 1, 2, \cdots, n)$$

when the symbols y and f in equations (1) are interpreted as row letters in the way apparently first introduced for differential equations by Peano.†

An interesting deduction from the theorems for a system of equations is the proof of the existence of a solution of a partial differential equation

$$F\left(x, y, z, \frac{\partial z}{\partial x}, \frac{\partial z}{\partial y}\right) = 0$$

which is not necessarily analytic in its five arguments, by means of the well-known theory of characteristic curves, as described in § 19.

---

* Vorlesungen über Variationsrechnung, page 179.

† "Intégration par séries des équations différentielles linéaires," *Mathematische Annalen*, vol. 32 (1888), p. 450.

§ 15. The Convergence Inequality

There is an inequality which is of frequent service in the existence proof of the following sections and which can be readily deduced from a simple preliminary theorem.

If u is a single-valued function of t with a well-defined forward derivative u' at each point of the interval $0 \leqq t \leqq t_1$, and if

$$|u'| < k|u| + l,$$

k and l being two positive constants, then u also satisfies the inequality

$$|u| \leqq |u_0|e^{kt} + \frac{l}{k}(e^{kt} - 1),$$

where u_0 is the initial value of u at $t = 0$.

Consider the function

$$v = |u_0|e^{kt} + \frac{l}{k}(e^{kt} - 1)$$

satisfying the differential equation

$$v' = kv + l$$

and having $|u_0|$ as its initial value. The value of u is never greater than that of v, since otherwise the difference $u - v$ would vanish and have a positive or vanishing forward derivative at some point. At a point where u and v are equal, however,

$$|u'| < k|u| + l = kv + l = v',$$

which is a contradiction. A similar argument shows that $-u$ is always less than v.

If u is a single-valued function of x with well-defined forward and backward derivatives at each point of an interval $x_0 \leqq x \leqq x_1$, and such that

$$|u'| < k|u| + l,$$

then, for any ξ and x in the interval, u also satisfies the inequality

$$(3) \qquad |u| \leqq |u(\xi)|e^{k|x-\xi|} + \frac{l}{k}(e^{k|x-\xi|} - 1).$$

This may be proved from the preceding paragraphs by putting $t = x - \xi$ for values of x greater than ξ, and $t = -x + \xi$ for values less than ξ.

§ 16. The Cauchy Polygons and their Convergence over a Limited Interval

It is proposed to consider a differential equation (1) for which the function $f(x, y)$ is continuous in the interior of a certain region R of the xy-plane, and such that the quotient

$$(4) \qquad \frac{f(x, y') - f(x, y)}{y' - y}$$

is finite when (x, y) and (x, y') lie in any closed region whose points are all interior to R.

A so-called Cauchy polygon for the equation (1) through a point (ξ, η) interior to R is defined by means of equations of the form

$$y_1 = \eta + f(\xi, \eta)(x_1 - \xi),$$
$$y_2 = y_1 + f(x_1, y_1)(x_2 - x_1),$$
$$\cdot \quad \cdot \quad \cdot \quad \cdot \quad \cdot \quad \cdot \quad \cdot$$
$$y = y_{n-1} + f(x_{n-1}, y_{n-1})(x - x_{n-1}).$$

The division points

$$\xi < x_1 < x_2 < \cdots$$

may be taken for convenience at equal distances δ from each other. Any value $x > \xi$ will lie on one of the intervals $x_{n-1}x_n$, and the polygon will either be well-defined for all such values, or else there will be a constant β such that for every x in the interval $\xi \leq x < \beta$ the points of the polygon are interior to R, while for $x = \beta$ the corresponding point (x, y) will be a point of the boundary of R. The polygon defined by the equations above may be denoted by $P_1(x)$, and the analogous one when the division points are distant $\delta/2^{n-1}$ from each other by $P_n(x)$.

A common interval $\xi \leq x \leq a$ for two functions $P(x)$, $Q(x)$ with respect to any region R may be defined as one over which both are interior to R, and one such that on any ordinate of the interval all the points between $(x, P(x))$ and $(x, Q(x))$ are also interior points of R.

Consider now a closed region R_1 interior to R and containing the point (ξ, η), and let m and k be two constants greater respectively than the absolute values of $f(x, y)$ and the quotient (4) in the region R_1. *If $l > 0$ is given in advance, the partitions for any two polygons $P(x)$, $Q(x)$ through (ξ, η) can be taken so small that*

$$(5) \qquad |P(x) - Q(x)| \leqq \frac{l}{k} \{e^{k|x - \xi|} - 1\}$$

for all values of x in any common interval of $P(x)$ and $Q(x)$ with respect to R_1. For at the point (x, y), where $y = P(x)$, the equation

$$P' = f(x, P) + \{f(x_{n-1}, y_{n-1}) - f(x, P)\} = f(x, P) + \rho$$

is satisfied by the forward and backward derivatives of the polygon P. On account of the continuity of $f(x, y)$ there exists for any l a constant μ such that

$$|x - x'| < \mu, \qquad |y - y'| < \mu$$

imply

$$|f(x, y) - f(x', y')| < l/2$$

whenever the points (x, y) and (x, y') are in R_1. If the subdivisions for $P(x)$ are taken less than μ and μ/m in length, it follows that on the polygon $P(x)$

$$|x - x_{n-1}| < \mu, \qquad |P(x) - y_{n-1}| < m|x - x_{n-1}| < \mu,$$

and hence the absolute value of ρ is less than $l/2$. Similarly $Q(x)$ satisfies an equation

$$Q' = f(x, Q) + \sigma,$$

where $|\sigma| < l/2$, provided that its intervals are less in length than μ and μ/m. The difference $P - Q$ has forward and backward derivatives which satisfy the relations ·

$$|P' - Q'| \leq |f(x, P) - f(x, Q)| + |\rho| + |\sigma|$$
$$< k|P - Q| + l,$$

and with the help of the lemma of § 15 the desired inequality follows at once, since P and Q have the same initial value η at $x = \xi$.

If $P(x)$ is a polygon and $Q(x)$ a solution of the differential equation, or if both are solutions, the same theorem evidently holds true, because then the function σ is identically zero, or else both ρ and σ vanish.

The polygons $P_n(x)$ all have a common interval. For take positive constants a and b such that the rectangle

(6) $$0 \leq x - \xi \leq a, \qquad |y - \eta| \leq b$$

is entirely within R, and consequently has two constants m and k analogous to those above for R_1. The portions of the polygons in the rectangle (6) all lie between the straight lines

$$y - \eta = \pm m(x - \xi),$$

since the slope of any side of any one of them is numerically less than m. It follows that each is certainly well defined and within the rectangle over an interval $\xi \leq x \leq a_1$, where a_1 is the smaller of a and b/m.

The sequence of polynomials $P_n(x)$ converges uniformly, on the interval $\xi \leq x \leq \xi + a_1$, to a function $y(x)$ which has a continuous derivative and satisfies the differential equation (1). *The curve $y = y(x)$ so defined is entirely within the region R.*

For take $\epsilon > 0$ arbitrarily, and l so small that

$$\frac{l}{k} \{e^{ka_1} - 1\} < \epsilon.$$

Then

$$|P_{n'}(x) - P_n(x)| < \frac{l}{k} \{e^{ka_1} - 1\} < \epsilon,$$

provided that the intervals $\delta/2^{n'-1}$ and $\delta/2^{n-1}$ are each less than the constant μ corresponding to l. Hence the sequence $P_n(x)$ converges uniformly to a continuous function $y(x)$ on the interval $\xi \leqq x \leqq \xi + a_1$.

The equations

$$P_n(x) \doteq \eta + \int_\xi^x P_n'(x)dx = \eta + \int_\xi^x \{f(x, P_n) + \rho_n\}dx$$

hold for every n, and the sequences $\{f(x, P_n)\}$ and $\{\rho_n\}$ approach uniformly the limits $f(x, y(x))$ and zero, respectively. Hence

$$y(x) = \eta + \int_\xi^x f(x, y(x))dx;$$

from which it follows by differentiation that $y(x)$ is a solution of the differential equation.

It is easy to show by means of the convergence inequality that there is only one continuous solution $y = y(x)$ of the differential equation (1) in the region R and passing through (ξ, η). For suppose there were another, $Y(x)$, distinct from $y(x)$ at a value $x' > \xi$. There would then be a value $\xi_1 < x'$ at which $y(\xi_1) = Y(\xi_1)$, and such that the two solutions would be distinct throughout the interval $\xi_1 < x \leqq x'$. In a neighborhood of the point of intersection (ξ_1, η_1) interior to R a relation

$$\left| \frac{d(Y - y)}{dx} \right| = |f(x, Y) - f(x, y)| < k|Y - y|$$

would be satisfied, and hence, from the convergence inequality (3),

$$|Y - y| \leqq 0.$$

This contradicts the hypothesis that $y(x)$ and $Y(x)$ are distinct throughout the interval $\xi_1 < x \leqq x'$.

§ 17. The Existence of a Solution Extending to the Boundary of the Region R

It has been proved in the preceding section that, on a certain interval $\xi \leq x \leq \xi + a_1$, the polygonal curves $y = P_n(x)$ converge uniformly to a continuous solution $y = y(x)$ of the differential equation (1) lying entirely within the region R. The interval for which the proof has been given may not be the largest one on which the sequence of polygons has this property. There will, however, be a number $\beta \geq \xi + a_1$, possibly infinity, with the property that on any interval $\xi \leq x \leq \beta_1$, where $\beta_1 < \beta$, the sequence of polygons converges uniformly to a continuous solution interior to R. A continuous curve $y = y(x)$ is thus defined which has a derivative and satisfies the differential equation for all values of x in the interval $\xi \leq x < \beta$.

As x approaches β the points $(x, y(x))$ of the solution can have no limit point (β, γ) interior to the region R.

If they did, there would be for any given ϵ a value $x' < \beta$ such that

$$|x' - \beta| < \epsilon, \qquad |y(x') - \gamma| < \frac{\epsilon}{2},$$

and an integer N such that, whenever $n \geq N$, the inequality

$$|P_n(x) - y(x)| < \frac{\epsilon}{2}$$

would hold for all values of x in the interval $\xi \leq x \leq x'$. At the value x' in particular

$$|P_n(x') - \gamma| \leq |P_n(x') - y(x')| + |y(x') - \gamma| < \epsilon;$$

so that for $n \geq N$ the points $(x', P_n(x'))$ would all lie in the ϵ-neighborhood of the point (β, γ). About the point (β, γ) as center a rectangle

$$|x - \beta| \leq A, \qquad |y - \gamma| \leq B$$

could be described entirely within the region R, and in the portion

R_1 of R which lay within the rectangle or within the region

$$\xi \leqq x \leqq x', \qquad y(x) - \epsilon \leqq y \leqq y(x) + \epsilon$$

the absolute values of $f(x, y)$ and the quotient (4) would be less than two constants m and k, respectively. It can be shown without great difficulty that every polygon $P_n(x)$ for $n \geqq N$ would be defined and lie within the region R for an interval extending beyond β at least a distance A_1, where A_1 is the smaller of the numbers A and $(B - \epsilon - m\epsilon)/m$. A proof similar to that of § 16 would then show that the polygons $P_n(x)$ converge uniformly to a continuous solution of equation (1) interior to R_1 over an interval $\xi \leqq x \leqq \beta + A_1$; and consequently β could not be the upper bound described above.

As x approaches β, therefore, the only limiting points of the solution $y = y(x)$ are at infinity or else are boundary points of the region R. If R is further a closed region, that is, one containing all of its limit points, then there is but one limit point for the curve $y = y(x)$ as x approaches β. For suppose (β, γ) to be a finite point in any neighborhood of which there are points on the curve. About (β, γ) a rectangle

$$(7) \qquad |x - \beta| \leqq A, \qquad |y - \gamma| \leqq B$$

can be chosen arbitrarily, and the points of R lying in it form a finite closed set in which $|f(x, y)|$ remains always less than a constant M. On the interval $\beta - A_1 < x < \beta$, where A_1 is the smaller of the numbers A and B/M, all the points of the curve $y = y(x)$ satisfy the inequality

$$(8) \qquad |y - \gamma| \leqq M(\beta - x).$$

For if (x', y') is any point of the curve in the rectangle (7) and also in an ϵ-neighborhood of the point (β, γ), then the inequality

$$|y - \gamma| \leqq |y' - y| + |y' - \gamma|$$
$$< M(x' - x) + \epsilon$$
$$< M(\beta - x) + \epsilon$$

must be satisfied by any preceding point $P(x, y)$ of the curve $y = y(x)$ for which the arc PP' is interior to the rectangle. It follows that the solution must lie interior to the rectangle and satisfy the last inequality, at least on an interval $x' - A_\epsilon < x < x'$, where A_ϵ is the smaller of $A - \epsilon$ and $(B - \epsilon)/M$. Hence the inequality (8) is also true on a properly chosen interval preceding $x = \beta$. It follows that as x approaches β there can be but one limit point for the curve $y = y(x)$, and this limit point is either at infinity or else is a boundary point of the region R

When the function $f(x, y)$ in the differential equation

$$\frac{dy}{dx} = f(x, y)$$

satisfies in a region R the conditions stated at the beginning of § 16, there exists through any interior point (ξ, η) of the region R one and but one continuous solution

(9) $y = \varphi(x, \xi, \eta)$

of the differential equation. This solution is defined and interior to R for all values of x interior to an interval

(10) $\alpha(\xi, \eta) < x < \beta(\xi, \eta),$

while as x approaches one of the end values α or β, the only limiting points of the solution are either at infinity or else on the boundary of R. If the region R is closed, then the solution has a unique finite or infinite limit point as x approaches α or β.

§ 18. The Continuity and Differentiability of the Solutions

It can be shown by methods due to Bendixon* that the function $\varphi(x, \xi, \eta)$ and its derivative $\varphi_x(x, \xi, \eta)$, whose existence has been proved in the preceding sections, are continuous in all three of their arguments, and if the function $f(x, y)$ has continuous first derivatives with respect to x and y in the interior of the region R,

---

* *Bulletin de la Société Mathématique de France*, vol. 24 (1896), p. 220.

then φ and φ_x have also continuous first derivatives with respect to all of their arguments.

The continuity at any set of values (x, ξ, η) for which (ξ, η) is in R and x satisfies the inequality (10) is provable with the help of the convergence inequality of § 15. For there will always be a region R_δ about the arc S of the solution (9) over the

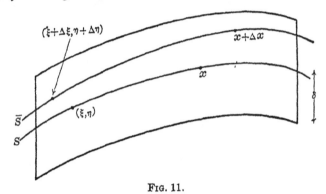

FIG. 11.

interval from ξ to x, of the kind symbolized in the figure, and so small that it lies entirely within the region R. If $(\xi+\Delta\xi, \eta+\Delta\eta)$ is any point in R, then the solution

$$(\tilde{S}) \qquad y = \varphi(x, \xi + \Delta\xi, \eta + \Delta\eta)$$

satisfies the inequality

$$(11) \quad |\varphi(\xi + \Delta\xi, \xi + \Delta\xi, \eta + \Delta\eta) - \varphi(\xi + \Delta\xi, \xi, \eta)|$$
$$= |\Delta\eta + \eta - \varphi(\xi + \Delta\xi, \xi, \eta)|$$
$$\leqq |\Delta\eta| + m|\Delta\xi|,$$

where m is the maximum of the absolute value of $f(x, y)$ in R_δ, on account of the relation

$$(12) \quad |\eta - \varphi(\xi + \Delta\xi, \xi, \eta)| = \left| \int_{\xi + \Delta\xi}^{\xi} \varphi(x, \xi, \eta)dx \right| \leqq m|\Delta\xi|.$$

Hence as long as $\tilde{S}$ remains within the region R_δ, it satisfies the convergence inequality

$$|\varphi(x,\ \xi + \Delta\xi,\ \eta + \Delta\eta) - \varphi(x,\ \xi,\ \eta)| \leqq \{|\Delta\eta| + m\overline{|\Delta\xi|}\}e^{|x-\xi-\Delta\xi|},$$

the initial values of the two solutions being taken at $x = \xi + \Delta\xi$. If $\Delta\xi$ and $\Delta\eta$ are sufficiently small the expression on the right is less than δ for all values of x belonging to the region R_δ, and hence $\bar{S}$ must be defined and interior to R_δ for all such values. Otherwise, for some interior value of x, it would attain one of the values $\varphi(x,\ \xi,\ \eta) = \delta$, which is seen to be impossible on account of the choice just made of $\Delta\xi$ and $\Delta\eta$.

Consider now the difference

$$|\varphi(x + \Delta x,\ \xi + \Delta\xi,\ \eta + \Delta\eta) - \varphi(x,\ \xi,\ \eta)|$$
$$\leqq |\varphi(x + \Delta x,\ \xi + \Delta\xi,\ \eta + \Delta\eta) - \varphi(x,\ \xi + \Delta\xi,\ \eta + \Delta\eta)|$$
$$+ |\varphi(x,\ \xi + \Delta\xi,\ \eta + \Delta\eta) - \varphi(x,\ \xi,\ \eta)|.$$

By a step similar to (12), and the inequality (11), it is seen to be less than

$$m|\Delta x| + \{|\Delta\eta| + m|\Delta\xi|\}e^{k|x-\xi-\Delta\xi|}$$

whenever $\Delta\xi$ and $\Delta\eta$ have been so chosen that S lies entirely in the region R_δ. Hence the continuity of $\varphi(x,\ \xi,\ \eta)$ is proved.

To prove the differentiability of φ with respect to ξ and η, assume that $f(x, y)$ has a continuous derivative f_y in the region R, and consider the same solutions S and $\bar{S}$ in the region R_δ. The difference of their ordinates satisfies the equation

$$\frac{d\Delta\varphi}{dx} = f(x,\ \varphi + \Delta\varphi) - f(x,\ \varphi) = A\Delta\varphi,$$

where, by Taylor's formula with the integral form of remainder,

$$A = \int_0^1 f_y(x,\ \varphi + u\Delta\varphi)du$$

is a continuous function of x, $\Delta\xi$, $\Delta\eta$, the values ξ, η being con-

8

sidered as constant for the moment. Hence

$$\Delta\varphi = ce^{\int_\xi^x A dx}.$$

When $\Delta\xi = 0$ or $\Delta\eta = 0$, the constant c has respectively the values

$$c = \Delta\varphi|_{x=\xi} = \varphi(\xi,\ \xi,\ \eta + \Delta\eta) - \varphi(\xi,\ \xi,\ \eta) = \Delta\eta,$$

$$c = \varphi(\xi,\ \xi + \Delta\xi,\ \eta) - \varphi(\xi,\ \xi,\ \eta) = \int_{\xi+\Delta\xi}^\xi f(x,\ \varphi + \Delta\varphi)dx$$

$$= - \Delta\xi f(\xi + \theta\Delta\xi,\ \varphi(\xi + \theta\Delta\xi,\ \xi + \Delta\xi,\ \eta)),$$

where $0 < \theta < 1$. Hence the quotients $\Delta\varphi/\Delta\xi, \Delta\varphi/\Delta\eta$ have well-defined limiting values

$$\frac{\partial\varphi}{\partial\xi} = - f(\xi,\ \eta)e^{\int_\xi^x f_y(x,\ \phi)dx}, \qquad \frac{\partial\varphi}{\partial\eta} = e^{\int^x f_y(x,\ \phi)dx}.$$

It may be remarked in conclusion that the theorems which have been proved in §§ 16–18 are true for systems of equations as well as for a single one.

§ 19. An Existence Theorem for a Partial Differential Equation of the First Order which is not Necessarily Analytic

Proofs have been given by Cauchy, Kowalewski, Darboux, and others for the theorem that in general there exists one and but one analytic surface

$$z = z(x,\ y)$$

which passes through an arbitrarily selected analytic curve C in the xy-space and, with the derivatives

$$p = \frac{\partial z}{\partial x}, \qquad q = \frac{\partial z}{\partial y},$$

satisfies a differential equation of the form

$$F(x,\ y,\ z,\ p,\ q) = 0,$$

where F is an analytic function of its five arguments. These proofs, however, say nothing about the solutions which may exist through a curve C whose defining functions are not expressible by means of power series; and they are not applicable when F itself has not this property. An existence proof is to be given below which is based upon much less restrictive assumptions on the functions F and the curve C. It involves the well-known theory of characteristic strips, which are solutions of a set of ordinary differential equations. If a one-parameter family of characteristic strips intersecting a given curve C is properly selected, it will generate a surface S which is a solution of the differential equation. The existence of the family and the differentiability of the surface depend, however, upon the existence and differentiability of the equations of the characteristic strips with respect to the initial values of the variables which they involve, that is, upon theorems similar to those which have been developed in the preceding sections.

Suppose that the function F is continuous and has continuous first and second derivatives in a certain region R of points (x, y, z, p, q). The differential equations satisfied by the characteristic strips have the form

(13)
$$\frac{dx}{du} = F_p, \quad \frac{dy}{du} = F_q, \quad \frac{dz}{du} = pF_p + qF_q,$$

$$\frac{dp}{du} = -F_x - pF_z, \quad \frac{dq}{dx} = -F_y - qF_z.$$

Through any initial values $(\xi, \eta, \zeta, \pi, \kappa)$ interior to R these equations have a solution with equations and initial conditions of the form

(14)
$$
\begin{aligned}
x &= x(u, \xi, \eta, \zeta, \pi, \kappa), & \xi &= x(0, \xi, \eta, \zeta, \pi, \kappa), \\
y &= y(u, \xi, \eta, \zeta, \pi, \kappa), & \eta &= y(0, \xi, \eta, \zeta, \pi, \kappa), \\
z &= z(u, \xi, \eta, \zeta, \pi, \kappa), & \zeta &= z(0, \xi, \eta, \zeta, \pi, \kappa), \\
p &= p(u, \xi, \eta, \zeta, \pi, \kappa), & \pi &= p(0, \xi, \eta, \zeta, \pi, \kappa), \\
q &= q(u, \xi, \eta, \zeta, \pi, \kappa), & \kappa &= q(0, \xi, \eta, \zeta, \pi, \kappa),
\end{aligned}
$$

and such that each of the functions on the left and its derivative for u are continuous and have continuous first derivatives in a region of values $(u, \xi, \eta, \zeta, \pi, \kappa)$ for which $(\xi, \eta, \zeta, \pi, \kappa)$ is a point interior to R and u lies in an interval, containing the value $u = 0$, of the form,

$$\alpha(\xi, \eta, \zeta, \pi, \kappa) < u < \beta(\xi, \eta, \zeta, \pi, \kappa).$$

The points (x, y, z, p, q) so defined are all interior to the region R.

Along the solution (14) the equations

(15) $$p x_u + q y_u - z_u = 0$$

(16) $$\frac{dF}{du} = F_x x_u + F_y y_u + F_z z_u + F_p p_u + F_q q_u = 0$$

are satisfied identically, so that the direction $p : q : - 1$ is always normal to the curve defined by the first three equations. Evidently if F vanishes at a single point of the strip, it will also vanish at every other point. The solutions (14) along which F vanishes are called characteristic strips, and any one of the strips (14) will surely be of this type if the initial condition

$$F(\xi, \eta, \zeta, \pi, \kappa) = 0$$

is satisfied.

Consider now a continuous and differentiable strip of elements

(17) $\quad x = \xi(v), \quad y = \eta(v), \quad z = \zeta(v), \quad p = \pi(v), \quad q = \kappa(v)$

$$(v_1 \leqq v \leqq v_2)$$

which lies in the interior of the region R and satisfies the conditions

(18) $\quad \begin{aligned} \pi \xi_v + \kappa \eta_v - \zeta_v &= 0, \\ F(\xi, \eta, \zeta, \pi, \kappa) &= 0, \end{aligned} \qquad \begin{vmatrix} F_\pi & \xi_v \\ F_\kappa & \eta_v \end{vmatrix} \neq 0,$

where the arguments in the derivatives of F are the same as those in the second equation. The first two of these conditions imply that the direction $\pi : \kappa : - 1$ is normal to the curve

(19) $$x = \xi(v), \quad y = \eta(v), \quad z = \zeta(v),$$

and that the curve and its strip of normals satisfy the differential equation. The third prevents the strip from being a so-called integral strip of the differential equation, through which there does not in general ·pass a unique integral surface without singularities. To make the situation simpler it will be supposed that the projection of the strip (17) in the xy-plane does not intersect itself.

When the functions (17) are substituted in the equations (14), a new system

(20)
$$x = X(u, v), \quad y = Y(u, v), \quad z = Z(u, v),$$
$$p = P(u, v), \quad q = Q(u, v)$$

with the initial conditions

(21)
$$\xi(v) = X(0, v), \quad \eta(v) = Y(0, v), \quad \zeta(v) = Z(0, v),$$
$$\pi(v) = P(0, v), \quad \kappa(v) = Q(0, v)$$

is determined. There is a region

(R_{uv}) $\qquad A \leqq u \leqq B, \quad v_1 \leqq v \leqq v_2,$

where A is a negative and B a positive constant, in which the functions (20) are continuous, have continuous first derivatives, and satisfy the relation

(22)
$$\begin{vmatrix} X_u & X_v \\ Y_u & Y_v \end{vmatrix} \neq 0.$$

For if M is the maximum of the absolute values of the functions on the right in the equations (13), for a closed ϵ-neighborhood of the points of the strip (17) in the interior of R, then the solutions (14) are defined at least over an interval $|u| \leqq \epsilon/M$, and the absolute values of A and B can be taken at least as great as this constant without disturbing the continuity properties desired for the functions (20) in the region R_{uv}. The condition .(22) is satisfied for the values $u = 0$, $v_1 \leqq v \leqq v_2$ because of the first two of equations (13) and the third of the relations (18); and the region R_{uv} can therefore be chosen so that the determinant is different from zero everywhere in it.

By an argument similar to that used in proving the theorem of § 4 it can be shown that A and B can be restricted still further, if necessary, so that no two distinct points (u', v'), (u'', v'') in the region R_{uv} define the same point (x, y) by means of equations (20). The boundary of the region R_{uv} is transformed then by the first two of equations (20) into a simply closed regular curve in the xy-plane which bounds a portion R_{xy} of the xy-plane. The equations establish furthermore a one-to-one correspondence between the points of R_{uv} and those of R_{xy}, and the functions

$$(23) \qquad u = u(x, y), \quad v = v(x, y)$$

so defined are continuous and have continuous first derivatives in R_{xy}. The others of the equations (20) define then three functions

$$(24) \qquad z = z(x, y), \quad p = p(x, y), \quad q = q(x, y)$$

which are also continuous and have continuous first derivatives in R_{xy}, and which with the values (23) for u and v satisfy the equations (20) identically in x, y.

The functions (20) satisfy the relations

$$(25) \qquad \begin{aligned} PX_u + QY_u - Z_u &= 0, \\ PX_v + QY_v - Z_v &= 0, \\ F(X, Y, Z, P, Q) &= 0, \end{aligned}$$

identically in u, v. The first and third of these follow at once from the equations (15), (16), the second of the equations (18), and (21). The expression

$$\Omega(u, v) = PX_v + QY_v - Z_v.$$

has the initial values

$$(26) \qquad \Omega(0, v) = \pi \xi_v + \kappa \eta_v - \zeta_v = 0,$$

which vanish on account of the first of equations (18).   Furthermore

$$\Omega_u = P_u X_v + Q_u Y_v + P X_{uv} + Q Y_{uv},$$

and from the first of equations (25),

$$0 = P_v X_u + Q_v Y_u + P X_{uv} + Q Y_{uv}.$$

By subtracting the last expression from that for Ω_u and using the equations (13) which the functions (20) satisfy, it follows that

$$Q_u = P_u X_v + Q_u Y_v - P_v X_u - Q_v Y_u = -\Omega F_z - \frac{\partial F}{\partial v},$$

in which the arguments of the derivatives of F are the functions (20).   Hence with the help of the third of equations (25) and the initial values (26),

$$\Omega_u = -\Omega F_z, \quad \Omega = \Omega(0, v) e^{\int_0^u F_z du!} = 0.$$

The single-valued function $z(x, y)$ defined above over the region R_{xy} has the derivatives

$$Z_x = \frac{\begin{vmatrix} Z_u & Y_u \\ Z_v & Y_v \end{vmatrix}}{\begin{vmatrix} X_u & Y_u \\ X_v & Y_v \end{vmatrix}} = p(x, y), \quad Z_y = \frac{\begin{vmatrix} X_u & Z_u \\ X_v & Z_v \end{vmatrix}}{\begin{vmatrix} X_u & Y_u \\ X_v & Y_v \end{vmatrix}} = q(x, y),$$

found by substituting the functions (23), (24) in the equations (20), differentiating the resulting identities, and applying. the first two of the relations (25).   It satisfies the differential equation $F = 0$ on account of the third of the equations (25).   Furthermore

$$x, \quad y, \quad z(x, y), \quad p(x, y), \quad q(x, y)$$

reduce to $\xi, \eta, \zeta, \pi, \kappa$ at any point of the strip (17), since at such a point $u(\xi, \eta) = 0$ and the relations (21) are satisfied.

It has been proved therefore that there is a single-valued

function

(27) $$z = z(x, y),$$

defined over a region R_{xy} of the xy-plane, which is continuous and has continuous first and second derivatives, contains the initial strip (17), and satisfies the differential equation $F = 0$.

There is no other surface

(28) $$z = z_1(x, y)$$

defined over the region R_{xy} and having these properties. If there were such a one, it would have to contain all of the points of the strips defined by equations (20). To prove this, suppose that (x', y', z', p', q') is an element belonging to one of the strips (20) for values (u', v'), and also to the surface (28). The equations

(29) $$\frac{dx}{du} = F_p(x, y, z_1, p_1, q_1), \quad \frac{dy}{du} = F_q(x, y, z_1, p_1, q_1),$$

where p_1 and q_1 are the derivatives of z_1, have a unique solution

(30) $$x = x_1(u), \quad y = y_1(u)$$

reducing to x', y' for the initial value $u = u'$ and defined over an interval $u' - \epsilon \leqq u \leqq u' + \epsilon$. The corresponding equations

(31) $$x = x_1(u), \quad y = y_1(u), \quad z = z_1(u),$$
$$p = p_1(u), \quad q = q_1(u),$$

found by substituting the functions (30) in z_1, p_1, q_1, define a characteristic strip. For on the surface (28) the equations

$$F_x + F_z p_1 + F_p r_1 + F_q s_1 = 0,$$
$$F_y + F_z q_1 + F_p s_1 + F_q t_1 = 0$$

are identities in x, y, where r_1, s_1, t_1 are the three second derivatives of $z_1(x, y)$. As a result of these identities and the equations (29),

$$\frac{dz_1}{du} = p_1 \frac{dx}{du} + q_1 \frac{dy}{du} = p_1 F_{p_1} + q_1 F_{q_1},$$

(32)
$$\frac{dp_1}{du} = r_1 \frac{dx}{du} + s_1 \frac{dy}{du} = - F_x - p_1 F_{z_1},$$

$$\frac{dq_1}{du} = s_1 \frac{dx}{du} + t_1 \frac{dy}{du} = - F_y - q_1 F_{z_1},$$

where the arguments of the derivatives of F are the functions (31). The equations (29) and (32) show that the strip (31) is a characteristic strip. Its initial element for $u = u'$ is (x', y', z', p', q'), the same as that for the strip (20) corresponding to $v = v'$. Hence the two must coincide on the interval $u' - \epsilon \leq u \leq u' + \epsilon$ on which both are defined.

The initial element (21) of any one of the strips (20) is by hypothesis on the surface (28). According to the last paragraph all of the elements of the strip in an interval $|u| \leq \epsilon$ must also lie on the surface, and it follows that there can be no upper bound except B for the values of u for which this is true. If $u' < B$ were such a limiting value, the element (x', y', z', p', q') corresponding to u' on the characteristic strip would also belong to the surface, on account of the continuity of $z_1(x, y)$ and its derivatives; and the interval of coincidence would therefore be necessarily longer than $0 \leq u < u'$.

For any point (x, y) in the region R_{xy} there is but one set of values (u, v) solving the first two of equations (20), and the corresponding value of z from the third equation belongs to both of the surfaces (27) and (28). The two surfaces must therefore coincide throughout.

Suppose now that an initial curve of the form (19) is given instead of the initial strip (17). If to any value v_0 defining a point (ξ_0, η_0, ζ_0) of the curve there corresponds a direction $\pi_0 : \kappa_0 : -1$ satisfying the relations (18), and such that $(\xi_0, \eta_0, \zeta_0, \pi_0, \kappa_0)$ is interior to R, then there will be a strip of elements of the form (17) along the curve containing these initial values for $v = v_0$. For the first two equations (18) have the solution (v_0, π_0, κ_0)

when their first members are regarded as functions of v, π, κ, and on account of the third relation (18) their functional determinant for π, κ does not vanish at these values. According to the fundamental theorem of § 1 there is therefore a pair of functions $\pi(v)$, $\kappa(v)$ defined over an interval $v_1 \leqq v \leqq v_2$ containing v_0 and satisfying, with $\xi(v)$, $\eta(v)$, $\zeta(v)$, the relations (18).

The results of the preceding paragraphs may be summarized as follows:

Suppose that

(C) $$x = \xi(v), \quad y = \eta(v), \quad z = \zeta(v)$$

is a continuous and differentiable curve, at some point (ξ_0, η_0, ζ_0) $= (\xi(v_0), \eta(v_0), \zeta(v_0))$ *of which there is a normal* $\pi_0 : \kappa_0 : -1$ *satisfying the equation*

$$F(\xi_0, \eta_0, \zeta_0, \pi_0, \kappa_0) = 0.$$

Suppose furthermore that

$$\begin{vmatrix} \xi_v(v_0) & F_p(\xi_0, \eta_0, \zeta_0, \pi_0, \kappa_0) \\ \eta_v(v_0) & F_q(\xi_0, \eta_0, \zeta_0, \pi_0, \kappa_0) \end{vmatrix} \neq 0,$$

and that the initial element $(\xi_0, \eta_0, \zeta_0, \pi_0, \kappa_0)$ *lies in a region R of points* (x, y, z, p, q) *in which F is continuous and has continuous first and second derivatives. Then there is a strip of the form*

(S) $$x = \xi(v), \quad y = \eta(v), \quad z = \zeta(v), \quad p = \pi(v), \quad q = \kappa(v)$$
$$(v_1 \leqq v \leqq v_2)$$

containing $(\xi_0, \eta_0, \zeta_0, \pi_0, \kappa_0)$ *for* $v = v_0$, *and such that all of its elements have the properties ascribed above to this initial one. If the projection C_{xy} of C in the xy-plane does not intersect itself, the characteristic strips of the differential equation*

$$F(x, y, z, p, q) = 0$$

which pass through the elements of S simply cover a region R_{xy} of the xy-plane and envelop a single-valued surface

$$z = z(x, y).$$

This surface is continuous and has continuous first and second derivatives in R_{xy}, contains the strip S, and satisfies the differential equation $F = 0$. There is no other surface over the region R_{xy} which has these properties.

CPSIA information can be obtained
at www.ICGtesting.com
Printed in the USA
BVOW10s1505180717

489611BV00009B/66/P

9 781164 003649